Pruning and Grafting

TIME
LIFE
BOOKS
®

Other Publications:

THE SEAFARERS

THE ENCYCLOPEDIA OF COLLECTIBLES

WORLD WAR II

THE GREAT CITIES

HOME REPAIR AND IMPROVEMENT

THE WORLD'S WILD PLACES

THE TIME-LIFE LIBRARY OF BOATING

HUMAN BEHAVIOR

THE ART OF SEWING

THE OLD WEST

THE EMERGENCE OF MAN

THE AMERICAN WILDERNESS

LIFE LIBRARY OF PHOTOGRAPHY

THIS FABULOUS CENTURY

FOODS OF THE WORLD

TIME-LIFE LIBRARY OF AMERICA

TIME-LIFE LIBRARY OF ART

GREAT AGES OF MAN

LIFE SCIENCE LIBRARY

THE LIFE HISTORY OF THE UNITED STATES

TIME READING PROGRAM

LIFE NATURE LIBRARY

LIFE WORLD LIBRARY

FAMILY LIBRARY:
   HOW THINGS WORK IN YOUR HOME
   THE TIME-LIFE BOOK OF THE FAMILY CAR
   THE TIME-LIFE FAMILY LEGAL GUIDE
   THE TIME-LIFE BOOK OF FAMILY FINANCE

Pruning and Grafting

by
OLIVER E. ALLEN

and

the Editors of TIME-LIFE BOOKS

TIME-LIFE BOOKS, ALEXANDRIA, VIRGINIA

Time-Life Books Inc.
is a wholly owned subsidiary of
TIME INCORPORATED

FOUNDER: Henry R. Luce 1898-1967

Editor-in-Chief: Hedley Donovan
Chairman of the Board: Andrew Heiskell
President: James R. Shepley
Vice Chairmen: Roy E. Larsen, Arthur Temple
Corporate Editors: Ralph Graves, Henry Anatole Grunwald

TIME-LIFE BOOKS INC.

MANAGING EDITOR: Jerry Korn
Executive Editor: David Maness
Assistant Managing Editors: Dale M. Brown, Martin Mann,
John Paul Porter
Art Director: Tom Suzuki
Chief of Research: David L. Harrison
Director of Photography: Robert G. Mason
Planning Director: Thomas Flaherty (acting)
Senior Text Editor: Diana Hirsh
Assistant Art Director: Arnold C. Holeywell
Assistant Chief of Research: Carolyn L. Sackett
Assistant Director of Photography: Dolores A. Littles

CHAIRMAN: Joan D. Manley
President: John D. McSweeney
Executive Vice Presidents: Carl G. Jaeger,
John Steven Maxwell, David J. Walsh
Vice Presidents: Peter G. Barnes (Comptroller),
Nicholas Benton (Public Relations), John L. Canova (Sales),
Herbert Sorkin (Production), Paul R. Stewart (Promotion)
Personnel Director: Beatrice T. Dobie
Consumer Affairs Director: Carol Flaumenhaft

THE TIME-LIFE ENCYCLOPEDIA OF GARDENING

EDITORIAL STAFF FOR PRUNING AND GRAFTING:
EDITOR: Robert M. Jones
Assistant Editors: Sarah Bennett Brash, Betsy Frankel
Text Editors: Margaret Fogarty, Bonnie Bohling Kreitler
Picture Editor: Neil Kagan
Designer: Albert Sherman
Staff Writers: Dalton Delan, Susan Perry, Reiko Uyeshima
Researchers: Diane Bohrer, Clarissa Myrick, Susan F.
Schneider, Betty Hughes Weatherley
Art Assistant: Santi José Acosta
Editorial Assistant: Maria Zacharias

EDITORIAL PRODUCTION
Production Editor: Douglas B. Graham
Operations Manager: Gennaro C. Esposito
Assistant Production Editor: Feliciano Madrid
Quality Control: Robert L. Young (director), James J. Cox
(assistant), Michael G. Wight (associate)
Art Coordinator: Anne B. Landry
Copy Staff: Susan B. Galloway (chief), Elizabeth Graham,
Lynn D. Green, Florence Keith, Celia Beattie
Picture Department: Barbara S. Simon
Traffic: Jeanne Potter

CORRESPONDENTS: Elisabeth Kraemer (Bonn); Margot
Hapgood, Dorothy Bacon (London); Susan Jonas, Lucy T.
Voulgaris (New York); Maria Vincenza Aloisi, Josephine du
Brusle (Paris); Ann Natanson (Rome). Valuable assistance
was also provided by: Leny Heinen (Bonn); Enid Farmer
(Lexington, Mass.); Anne Angus (London); Diane Asselin
(Los Angeles); Carolyn T. Chubet, Miriam Hsia (New York).
The editors are indebted to Ellen Cramer, Angela Goodman,
Michael McTwigan, Jane Opper, Maggie Oster, Lee Lorick
Prina, Karen Solit, Lyn Stallworth, Sandra Streepey, Anne
Weber, writers, for their help with this book.

THE AUTHOR: Oliver E. Allen is a former staff member of LIFE magazine and of TIME-LIFE BOOKS, where he served as Editor of the LIFE World Library and the TIME-LIFE Library of America. He also wrote *Decorating With Plants* for The TIME-LIFE Encyclopedia of Gardening.

CONSULTANTS: James Underwood Crockett, author of 13 of the volumes in the Encyclopedia, co-author of two additional volumes and consultant on other books in the series, has been a lover of the earth and its good things since his boyhood on a Massachusetts fruit farm. He was graduated from the Stockbridge School of Agriculture at the University of Massachusetts and has worked ever since in horticulture. A perennial contributor to leading gardening magazines, he also writes a monthly bulletin, "Flowery Talks," that is widely distributed through retail florists. His television program, *Crockett's Victory Garden,* shown all over the United States, has won millions of converts to the Crockett approach to growing things. Dr. Miklos Faust is Chief of the Fruit Laboratory at the U.S. Department of Agriculture Agricultural Research Center, Beltsville, Md. Paul F. Frese is a lecturer, writer and horticultural consultant and edited the *Pruning Handbook* of the Brooklyn Botanic Garden. Dr. Conrad B. Link is Professor of Horticulture at the University of Maryland, College Park. Edwin F. Steffek is Editor Emeritus of *Horticulture* magazine and author of *The Pruning Manual.*

THE COVER: Once an opaque mass of green boughs, a 75-year-old Sargent's weeping hemlock has been heavily pruned to reveal, against a wall, the interior drama of its trunk and limbs. Clouds of foliage at the end of each branch create the bonsai-like illusion of an ancient wind-swept tree. In winter, as these evergreen tufts collect snow, the tree appears to be laden with white blossoms.

Library of Congress Cataloging in Publication Data
Allen, Oliver E., 1922
    Pruning and grafting.
    (The Time-Life encyclopedia of gardening ; v.23)
    Bibliography: p.
    Includes index.
    1. Pruning. 2. Grafting. I. Time-Life Books. II. Title
SB125.A44   635.9′1′541   78-15795
ISBN 0-8094-2635-8
ISBN 0-8094-2634-X lib. bdg.

CONTENTS

The when and why of pruning 1

It was a sunny morning in early April when Henry Upton found time to do some pruning. His shrubbery looked a bit overgrown—he hadn't touched it in several years—and the trees needed thinning. So he started cutting and slicing in a grand manner. In the course of his work, he lopped several branches from a pair of maples and a flowering dogwood, and he carefully thinned his azaleas and rhododendrons. For good measure, he sharply cut back an overgrown forsythia bush. He felt a great sense of accomplishment.

Several weeks later he became aware that something was wrong. The maples and the dogwood were oozing quantities of sap from their wounds. Worse, the azaleas, rhododendrons and forsythia showed no sign of bloom. His yard, usually a springtime show place, looked dreary and forlorn.

From a friend, Henry learned that his instincts were sound but his timing was faulty. Maples and dogwoods, he learned, bleed sap—harmlessly, but unattractively—when they are cut in the spring; they should be pruned in the summer when the sap is not rising. The shrubs were early bloomers that formed flower buds the previous year; in his zeal, Henry had removed practically all of the potential blossoms. Those plants should have been pruned immediately after they finished blooming.

Many gardeners have duplicated Henry Upton's experience, and needlessly, for it is easy to determine the preferred pruning season for different kinds of plants. Others ignore the sensible rules of pruning, attacking trees and shrubs as if trying to get even with them, shearing with hedge clippers to produce a horticultural crew cut that robs each plant of its individuality. Still others are loath to cut plants at all, feeling that to do so contravenes nature.

But those who prune efficiently and perceptively are rewarded with healthier and better looking plants, shrubs and trees. "Good pruning dresses up your garden instantly," says one professional. "It

The fiery glow of a sunset silhouettes the branches of two coast live oaks, pruned to open up their dense foliage to sunlight and air. The trees will now need only light pruning to remove barren branches.

is much like changing the summer slip covers in your living room." Its practitioners also discover that pruning, along with the allied practice of grafting, is one of the most creative aspects of gardening, for it is the key to controlling the sizes and shapes of plants or their flowers and fruits.

As to the argument that it is against nature, a stroll through any patch of woods provides the answer. Nature itself eliminates many branches, but in a ruthless, haphazard and messy way. Some branches die but stay on the trees, inviting disease and insect attacks. Branches that grow at weak angles snap off in storms, leaving ugly rips in the trunks. When branches fall they may crush shrubs below. Plants compete for light, the strong ones reaching up while the weak are thrust aside. Such a survival-of-the-fittest competition is inappropriate for the garden. You are more likely to want to develop the natural shape that trees and shrubs have when they are at their best. To do this, you will need to remove extraneous growth before it becomes large enough to be troublesome.

WHAT PRUNING CAN ACHIEVE   Someone once remarked that pruning is not difficult—all you do is cut off a branch. But there is more to it than that. Your first objective is to maintain the health of your plants by keeping them free of the three Ds: dead, damaged or diseased branches. But beyond this are several other goals.

You prune to shape a plant as it grows, especially when it is young, making it shorter, thinner or bushier, according to the purpose you have in mind. Fuchsias, for example, can be trained to sprawl or to stand upright like trees. Hemlocks can stand alone as ornamentals or grow close together in a hedge line. Forsythias can be made thick enough to block a view, even in the winter, or can be induced to arch gracefully along a slope.

Frequently a plant becomes too large for the space originally allotted to it—as with an overgrown rhododendron or hydrangea in a foundation planting, or even a burgeoning indoor plant that threatens to crowd you out of your living room. In this situation, you must prune a different way to keep it in bounds.

When a plant becomes old and leggy, you can rejuvenate it, eliminating aged growth so young shoots can take over. Or you can prune selectively to correct some defect in a plant—thinning it so more light reaches the inner branches, perhaps, or eliminating a branch that rubs another, or cutting off a branch growing at such a narrow angle that it is vulnerable to breakage, or cutting back enough branches so the plant's top and its roots are balanced.

If your plant is of a flowering or fruiting variety, you can prune to achieve the best conditions for copious blossoming and fruiting.

Or, if you like, you can reverse the process to curb flowering while stimulating foliage growth.

Finally, you can use your pruning skills to fashion novel plant forms. You may want to try the sculptural shapes known as topiary, or the graceful two-dimensional patterns achieved by the technique called espalier, or the deliberate dwarfing of certain kinds of trees and shrubs, or other innovations.

But keep in mind that your most urgent pruning objectives will be to train a young tree or shrub by taking out undesirable branches and cutting others back so the plant becomes wider or taller, to thin an older plant by removing some branches entirely to let in more light, or to reduce the size of a large plant by removing portions of branches at points where there are weak replacement shoots.

In order to prune effectively and efficiently, you need to know why cutting has the effect that it does on a plant. When you know how plants grow, how they form new tissue and how they respond to light and water and the change of seasons, the logic of pruning becomes clear. Most plants grow in more or less the same way; what you accomplish when you pinch back a window-sill begonia is not unrelated to what happens when you prune an oak. The differences are mostly a matter of scale.

One way to understand how shrubs and trees function is to visualize each plant as containing interconnected networks that constantly convey food and energy from one part to another. At the bottom, in the soil, are roots that gather water and nutrients to send aloft. At the top of the plant are the leaves (perhaps in the form of needles) that carry out a complex function of using light to convert water, nutrients and carbon dioxide from the air into the sugars that provide the tree or shrub with energy. In between, the trunk, stem, branches or shoots pass these substances back and forth to keep the plant alive and growing.

MANIPULATING HORMONES

This entire system is controlled by hormones. When you prune, you tinker with this hormonal system. This influences the way the plant will grow. Above all, you tinker with the buds at the tips of branches or stalks. Pruning removes some buds and in doing so forces others to grow. This growth into shoots is called primary growth; it extends the plant lengthwise so it can reach higher into the air or farther to each side. (The kind of growth that increases the girth of trunk or branches is called secondary growth.)

Buds are extraordinarily complicated. If you open one and examine it under a microscope, you will find that it contains everything needed to create leaves and branches or, in some cases, flowers and fruit. The most important buds are those that grow at the tips of

Basic anatomy for the surgeon

In a real sense, pruning and grafting are surgery. Like any good surgeon, the gardener who understands the anatomy of what he is cutting does the best work. All woody plants, both trees and shrubs, are similar in structure. A skeletal system supports a fibrous network of cells that circulate water, sugar and other nutrients up and down, while an outer skin fends off infection and predators.

Paradoxically, the heart of a plant's anatomy is not at the heart of the plant but in a thin layer of tissue near the outside of every stem, branch and trunk. This is the cambium, only one cell thick and invisible to the naked eye. From it come all the plant's other cells—bark, wood, buds, leaves and flowers. If the cambium is severed by a careless knife, everything beyond that point dies.

OUTER BARK
Covering the plant like a sleeve, this layer of old, dead cells shields the living cells from cold, wind and rain and protects them from pests and diseases. When you prune or graft, take care not to let bark peel beyond the edge of the cut, since this exposes tender inner bark.

INNER BARK
Lying next to the cambium layer and composed of young, active cells, the inner bark, called the phloem, carries sugars manufactured by the leaves and distributes them through the plant. When the outer bark is thin, as on stems and new shoots, sunlight induces chlorophyll to form in the inner bark, turning the bark green. During winter pruning, this green layer indicates that a branch is alive.

CAMBIUM
Most active in early spring when soil warms and sap rises from the roots, the cambium cells divide, inside and outside, thus increasing the plant's girth. The new inside cells become wood tissue, those on the outside become bark. In grafting, the cambium unites the graft.

SAPWOOD
The ever-thickening sapwood layer inside the cambium carries water and nutrients from roots to leaves. Each year's growth becomes an annual ring, marked by denser tissue as the seasonal rate of growth slows. Vascular rays run laterally through this sapwood, carrying water and nutrients.

HEARTWOOD
As sapwood cells die, they harden into dark heartwood, providing the plant with a stiff internal support.

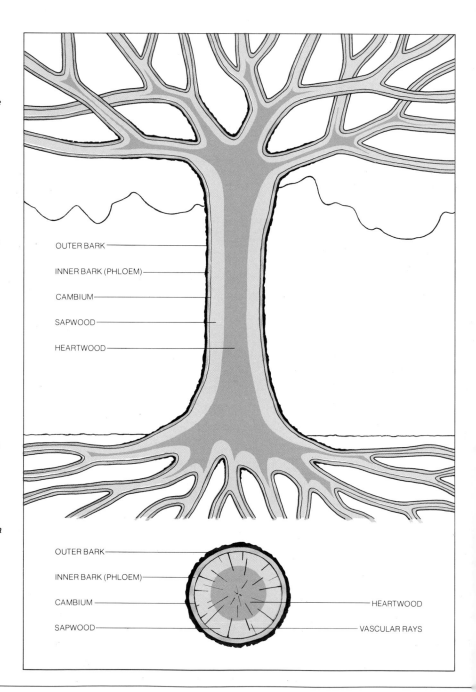

OUTER BARK
INNER BARK (PHLOEM)
CAMBIUM
SAPWOOD
HEARTWOOD

OUTER BARK
INNER BARK (PHLOEM)
CAMBIUM
SAPWOOD
HEARTWOOD
VASCULAR RAYS

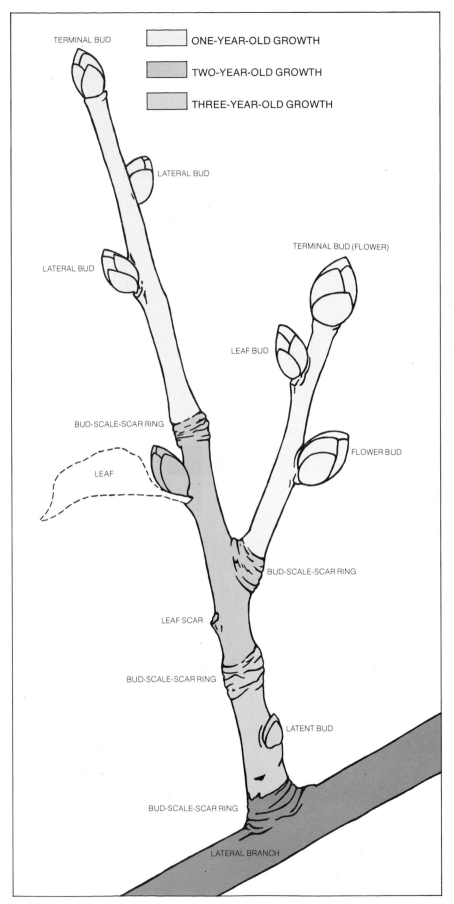

TERMINAL BUD

ONE-YEAR-OLD GROWTH

TWO-YEAR-OLD GROWTH

THREE-YEAR-OLD GROWTH

LATERAL BUD

LATERAL BUD

TERMINAL BUD (FLOWER)

LEAF BUD

BUD-SCALE-SCAR RING

LEAF

FLOWER BUD

BUD-SCALE-SCAR RING

LEAF SCAR

BUD-SCALE-SCAR RING

LATENT BUD

BUD-SCALE-SCAR RING

LATERAL BRANCH

TELLING THE BUDS APART

Buds are the embryonic organs of plants, containing the beginnings of stems, leaves and flowers. From undifferentiated cells tightly packed inside protective coverings, or bud scales, they gradually develop into three kinds of buds according to location—terminal, lateral or latent, and two kinds of buds according to function—vegetative buds that form leaves and stems, and flower buds. The plant also has the ability to form emergency buds where none exist, called adventitious buds.

TERMINAL BUD
Located at the very tip of any stem or branch, the terminal bud controls the ultimate length or height of that branch. A hormone within this bud inhibits the growth of adjacent buds along the same stem. If the terminal bud is injured or removed, some of these adjacent buds will grow.

LATERAL BUD
Formed at the thickened node where a leaf joins a stem, most lateral buds develop into shoots and leaves in their first growing season, but a few may remain dormant for a year or more. Every node contains cell tissue capable of healing a cut.

LATENT BUD
A latent bud is a dormant lateral bud that can remain inactive for many years. But if an accident or a deliberate pruning cut removes a nearby shoot or branch, the latent bud will be stimulated to develop.

ADVENTITIOUS BUD
This bud does not come into existence unless there is a major injury to a plant and no nearby latent buds are present to take over. Whereas other buds are formed in specific locations from specialized cells, adventitious buds are formed wherever needed.

LEAF AND FLOWER BUDS
All young buds are vegetative buds and are thin and green. But an unknown stimulus transforms some into bigger, fatter, grayish flower buds.

BUD-SCALE-SCAR RING
As terminal buds open and grow, they leave behind a series of wrinkle-like rings called bud-scale scars. By counting the rings, you can tell the age of a stem or branch. If you cut at a ring, the cell tissue present there will quickly divide to cover the wound.

Thinning, one of the most basic pruning techniques, opens a plant to sunlight and air: it relieves crowding and allows space for new shoots and leaves to develop. To thin is to remove an entire branch right back to the larger branch from which it grows (upper red bar). Trees are sometimes thinned back to the trunk itself, and in the case of some woody shrubs, thinning involves cutting back some of the stems to ground level (lower red bar). Thinning is often undertaken during the summer.

branches. They are called terminal buds, and they receive the most energy as they seek sunlight. Slightly less energy goes to the lateral buds down the side of the branch or shoot, so they grow at a slightly slower pace. These are sometimes called axillary buds because they often appear at the axil, the joint where a leaf joins the stem. These lateral buds may appear in clusters, in pairs (one on each side of a branch or stem) or, as in rose bushes, alternating from one side of the stem to the other.

Wherever lateral buds appear, the branch is likely to be slightly thickened; at that point it contains tissues that are capable of growing when they are stimulated to do so. This thickened area is called a node. The spaces between nodes, which do not have such tissues, are called internodes.

Just behind each lateral bud there is likely to be a latent bud—one that may never break through the surface but is ready to begin growing if something happens to destroy the terminal and lateral buds. And finally, what are called adventitious buds may be created from scratch by the plant when they are needed. These can appear almost anywhere on the plant, even on the roots, but only under severe stimulus—great damage to the tree, perhaps a disease, or some extremely hard pruning.

The terminal bud—the one at the very tip of any shoot or branch—is the source of auxin, the hormone formation that inhibits the growth of the side buds. This phenomenon is called apical dominance, since the terminal bud is at the apex of that shoot. If you remove the terminal bud, the absence of its dominance permits lateral buds to produce shoots that grow rapidly until a new terminal bud establishes dominance. If both terminal and lateral shoots are removed, latent and even adventitious buds will start to grow, even on thick branches or the trunk, and the tree or shrub will develop a hairy look. Most of the time, such shoots are undesirable and are commonly called suckers or water sprouts.

When you remove a terminal shoot and its bud just above a node, the first reaction is likely to take place just below the cut, in the lateral bud at that very node. Plants differ in this regard, however. Some, like the elm tree, will start new growth at several points along that branch. A palm has no lateral buds at all; if its terminal bud is removed, a palm will die. But most familiar trees and shrubs will continue to grow with their energies redirected. And it is just as well that all the buds do not put out shoots at once, for the tree would never be able to support them all and it would forfeit its defense against adversity.

Experts speak of two basic pruning techniques: thinning and

cutting (or heading) back. When you thin a plant, you remove an entire branch or stalk, cutting it off at the next larger limb, at the trunk or main stem, or even at ground level. The effect of thinning is to open up the center of a plant by reducing the number of branches without stimulating more growth; this is often done to admit more light into the middle. But when you cut back, you remove much less, only part of a shoot, cutting at a bud or node not far from the tip. Cutting back removes apical dominance, stimulating side growth and thickening the plant with new foliage.

The severity of pruning usually determines a plant's response. The more old growth you remove, the more new growth the plant will produce. Some plants thrive on radical surgery: if you cut hills-of-snow hydrangea branches 4 to 6 inches from the ground each year, the plant will become 3 feet tall and develop massive flower clusters. Many kinds of roses demand severe pruning (*page 49*). With trees, cutting back is usually done only in the first three to five years after planting; when a tree is fully developed, usually only thinning cuts are used.

Any time a terminal shoot is removed, the plant will try to replace it. Several of the buds at nodes below the cut will begin growing vertically rather than to the side. The one at the node where the cut was made is most likely to take charge at the apex. If you inspect a tree or shrub several months after removing a terminal shoot, the only evidence that something happened is likely to be a slight irregularity in the branch at the point of pruning.

In many cases you can steer a plant's growth by carefully

A TERMINAL TAKEOVER

CUTTING BACK TO SPUR NEW GROWTH

1. *Cutting back, sometimes called heading back, is the basic pruning technique for stimulating new growth to thicken a plant. The cut removes only part of a branch, back to a selected bud. With the branch's terminal bud eliminated, so is a hormone that inhibited growth of nearby buds.*

2. *From buds nearest the cut, new shoots develop, each in time developing its own terminal bud and side buds.*

3. *The topmost new shoot fills the space left by the old tip, while the other shoots continue to grow and leaf out, filling the end of the branch with additional foliage.*

choosing the bud and node where you cut. An outward-facing bud will produce a branch that grows outward; by choosing it you can open up the center of a plant and increase its spread. But if you want to thicken a scraggly, rangy shrub, cut the branch just beyond an inward-facing bud.

There is a reverse alternative to such cutting back. If you want a plant or tree to grow taller or wider rather than have heavier, bushier foliage, removing side buds or shoots will allow more nutrients to be transported to the terminal tip.

IDENTIFYING FLOWER BUDS

In the case of a flowering tree or shrub, your decision on where to cut may hinge on whether or not the bud is vegetative—one that will produce a leaf or branch instead of a flower. It is usually easy to tell the buds apart. Flower buds are almost always larger and fatter than vegetative ones. Many flowering trees and shrubs form both kinds of buds in one summer to develop the next spring; in the intervening winter you can see both kinds of buds on a single stalk or twig. On peach trees, buds often grow in clusters of three. The center one will be vegetative, putting out a leafy shoot, while the other two will bear flowers and perhaps eventually produce fruit.

The only kind of pruning in which you need not concern yourself with buds, nodes or branching points is not strictly pruning

BRACING FOR GOOD SPACING

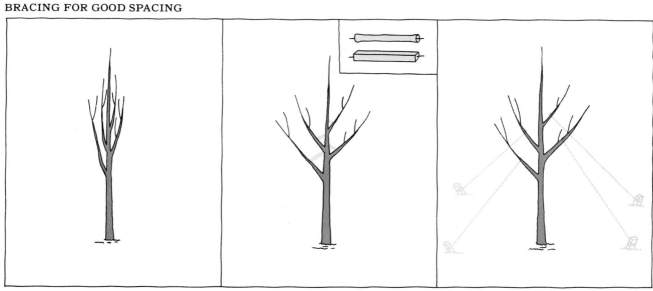

A young tree's pencil-thin branches tend to grow upright and crowded together. Bracing is the best technique for spacing them evenly and widening narrow crotches; pruning deprives the tree of food-manufacturing leaves, slowing growth.

Make braces from wood scraps 12 to 18 inches long. Drive a nail partway into each end and clip off the head. Or run wire through bamboo sections so that it extends ¼ inch at each end (inset). Wedge the braces between branches. Leave for several months.

Branches can also be spaced with taut string. Determine which branches need repositioning, drive stakes into the ground and tie cotton string to each branch. Pull the branch down and tie the string to a stake. Leave the string in place about a year.

at all. When you shave a formal hedge or a piece of topiary sculpture *(Chapter 4),* you are shearing, not pruning. Familiar hedge plants like privet, boxwood and yew have so many responsive latent buds that they will grow almost anywhere you cut, quickly covering any shorn expanse with fresh new greenery. But only plants that can easily stand rough treatment should be used for geometrically shaped hedges. Other plants can be set in hedgerows, of course, but you will have to keep them under control with selective pruning rather than shearing.

A key fact to remember about both pruning and shearing is that only a bud can make a branch longer. A branch grows only at its tip, not by stretching out. While a tree keeps growing taller, any one branch remains at the height where it started. If a branch is growing out of a trunk 6 feet above the ground, it will still be 6 feet up 25 years from now, though it will be thicker.

Roots, however, do not grow from buds. They expand their cores lengthwise at growing tips, and they take in moisture and nutrients mainly through tiny root hairs. The overall size of the root structure usually equals that of the canopy of the tree or shrub, but it is shallower and wider.

HOW ROOTS FUNCTION

Because roots and top should be in balance, you may have to prune the top occasionally to maintain that balance. But you can also deliberately upset the balance to affect your plant in some special way. Root-pruning is frequently used to limit a plant's growth, since it decreases the supply of nutrients that reaches the top. If you want to keep a house plant at a certain size, for example, remove it from its pot and snip off a portion of the roots all around. You can also root-prune a shrub by pushing a sharp spade into the ground at intervals around the base, severing the growing tips of some but not all of the plant's feeder roots.

THE SENSITIVE CAMBIUM

An astonishing thing about a plant is that you can invigorate it by cutting off buds, shoots or even branches, but if you injure the layer of sensitive tissue called the cambium, you may kill a limb or shoot, or even an entire plant. It is as if losing a finger would make you healthier but skinning your knee would be fatal. The moist, microscopically thin layer of tissue is surprisingly delicate. It produces the side cells adjacent to it that convey a plant's food up and down and make the limbs grow thicker. The cambium also controls the way pruning cuts will heal, and it is the key to all successful grafting *(Chapter 3).*

Inside the bark of a woody plant, even the green bark of a slender twig, are three layers of tissue that are critical to the plant's health. The innermost is the sapwood, through which water and nutrients rise from the roots up into the plant. This layer changes color during the year, creating the annual rings that you can count on a felled tree to determine its age. Next comes the cambium. Outside the cambium, just under the bark, is a layer of inner bark called the phloem (pronounced flow-em); it conveys the food manufactured by the leaves down into the tree.

So vital are these layers that even if the entire inside of a tree is hollowed out, top to bottom, the tree will continue to live. But a thin wire stretched tight around that tree will kill it if the wire cuts through the phloem and cambium as the tree grows. This process is called girdling. A carelessly planted tree can be strangled by one of its own roots wrapped around it or by uncut string holding its burlap root-ball wrapping; a branch can be killed by the tight wire of a plant label left in place. Girdling is so deadly that, to prevent it, wire braces used to hold a young tree erect must be loose and covered with soft rubber tubing. When a critical amount of damage is done, perhaps by hungry rodents or a lawn mower, a prized tree can sometimes be saved with a special grafting technique used to bridge the gap in the vital layers (page 69).

CARE OF EVERGREENS

Needled evergreens differ somewhat from deciduous trees in the way they grow. They grow more slowly, for example, and they often need no pruning at all. But they may sustain damage requiring repair, or you may want to prune to keep them a certain size.

The so-called fine-foliage conifers—hemlocks, yews, arborvitae and junipers among others—will withstand quite radical cutting back and can even be sheared, especially when they are young. But much more conservative handling is demanded by the long-needled conifers—pines, spruces and firs—which do not readily form new branches. These do, however, put out new growth each spring in the form of extended candle-like buds from which new needles will appear. If you cut or pinch off a portion of each candle, the tree's growth will be limited; if you remove just the center candle, side candles will grow and branch, thickening the tree.

Apical dominance is so pronounced in the long-needled conifers that side branches near the top are always small, getting progressively longer farther down. These evergreens do not readily replace a top leader. If one is injured or lost, you can replace it by tying a nearby side branch into a vertical position, using a stick as a splint held on with cloth or soft twine.

Almost everyone agrees that needled evergreens should be pruned in the spring, just before or during their period of strongest growth. But there is no such unanimity about when to do other pruning. In fact, many experts say plants can be pruned at any time—"whenever your knife is sharp" or "on all but the 52 days of the year that fall on Sundays." Indeed, there is some pruning work that can be done all year long if you have a wide enough variety of plants. Certainly any of the three Ds—dead, damaged or diseased wood—should be removed as soon as it is spotted. Trees that bleed, as Henry Upton discovered, look better if trimmed in the summer, even though bleeding has no ill effect. Beyond that, when you prune will be dictated by each plant's annual growth cycle.

In general, when plants are in full growth during the summer, they use all of their energy increasing their size and producing flowers or fruit. As autumn approaches, new buds form, flower buds become differentiated, and the food manufactured begins to go not into growth but into storage, mostly in the roots. In winter the plant is dormant, but as the temperature rises in early spring, the sap also begins to rise, stored food begins to flow and growth resumes, with limbs elongating and buds bursting open. Until the leaves are fully opened, the plant continues to live on stored food. Then the new leaves take over.

When you prune in winter or early spring, you remove some of the buds. If some of these are flower buds, you may not want to lose them. Otherwise, you have simply reduced the number of growing points that will divide up the springtime burst of energy. There will be stronger growth in the branches that remain, especially in a young tree. "If you want wood, prune in winter," goes an old saying. Indeed, the response to such pruning may be so strong that the remaining branches cannot absorb all of the available energy. The tree will put out new branches, perhaps including a crop of suckers around the base. (They should be removed as they appear.)

Conversely, when you prune in summer, you are removing some leaves that would otherwise be manufacturing food. With the leaf loss there probably will be a decrease in root growth as the tree compensates. This reduces the amount of food that can be stored and released the following spring, thus curbing the plant's growth. So summer pruning is better if it is your intention to hold down the size of a tree or shrub.

However, you incur less risk with winter pruning, especially if you do it early in the winter and spread major pruning over two or three years. Summer thinning should be done either early enough so the plant has time to recover or after growth has ceased for the season. Avoid cutting back in the summer; it forces the tree or shrub to produce new wood that does not have time to develop storage material; it will not be reliably hardy and may be killed by even relatively mild winter weather.

CHOOSING A TIME TO PRUNE

With flowering shrubs, your special concern is their time of bloom. Those that set their flower buds one year for bloom the next, such as azaleas and forsythia, should not be touched with pruning shears until they have finished blooming (unless you are willing to forgo the bloom, of course). But then, as with trees, it is a good idea to complete the pruning as quickly as possible after flowering has ended. That gives the plants time to set new flower buds for the next blooming season. But shrubs that blossom in the summer or fall—for example, buddleia, tamarix and summer-flowering spirea—can be pruned either during their period of winter dormancy or in the early spring. They initiate flower buds on the new wood that grows during the season of bloom.

In mild climates where the temperature does not vary much between seasons, different rules govern pruning. In some parts of the United States there are only two seasons, wet and dry. In such areas the best time to prune is just after the plants have emerged from dormancy and started new growth, usually toward the end of the dry season.

House plants undergo little temperature or moisture change during the year. They may be pruned lightly at any time, but if you want to stimulate strong new growth, prune during the most active growing season, probably in spring and early summer. House plants that flower but once a year, like spring-flowering shrubs, should be pruned just after they have blossomed.

BETTER GROWTH, NOT MORE

In considering the general effects of pruning, it is important to note that pruning alone will not increase the total growth of any plant, shrub or tree. Only improved growing conditions or better nutrition will do that. What pruning does is to redirect energy within the plant so it will assume a different shape, produce flowers or fruit in a different manner, or renew itself more efficiently. The actual pruning has the over-all effect of limiting and even slowing the plant's total growth. What remains after judicious cutting simply grows better. Pruning is most crucial for a young plant or tree, for then you can guide it toward an ideal shape or function. When it reaches maturity, you will prune only to maintain or renew it.

Once a shade tree grows past 20 or 30 feet, give up pruning it yourself and let the professionals take over. Working high in a large tree is difficult and dangerous for an amateur; it should be left to experts with hydraulic lifts and power equipment, or to skilled tree climbers. As a rule of thumb, cut only what you can reach from a stepladder, using an ordinary pair of pruning or lopping shears, or what you can reach from the ground with a pole pruner (page 37). Above all, never take a chain saw aloft. Use it only on the ground for

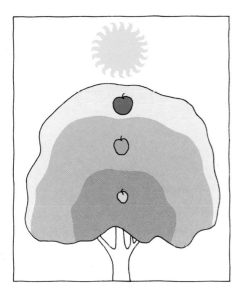

BRIGHTEST AT THE TOP
An unpruned tree may cast such dense shade on its own lower branches that they produce small leaves, few blossoms, or fruit of poor quality. A U.S. Department of Agriculture study revealed that the sunny upper third of an unpruned apple tree bears large, bright red fruit. In the middle section fruit is smaller and paler, and the heavily shaded bottom yields only a scattering of nubbins.

cutting up limbs that have been manually sawed off. Many observers have noted that a tree surgeon does not charge nearly as much as the one at the hospital.

As to when to call in the experts, cast a critical eye on your large trees from time to time to see if they are cluttered with dead branches or show signs of disease. Have large dead limbs removed as soon as possible for safety's sake. If a number of branches and twigs come down in a thunderstorm, that is a sign the tree needs work. Watch for crossed branches that rub each other, and branches so weighed down with foliage that they sag. See if light penetrates the center of the tree; as one professional observed, "If you do not see any sky when you look up or any sunlight on the ground below, the tree needs thinning."

Meanwhile, you will have plenty to keep you occupied among the smaller trees, shrubs and other plants in your yard. Keep a plan in mind at all times, and never cut anything without knowing why. The ideal pruner never shows his tracks. Unless you are deliberately creating an artificial shape like a formal hedge, an espaliered tree or a piece of topiary sculpture, your pruned plant should look as though nothing has happened to it. Prune to the plant's natural shape and let it reveal its own beauty.

Finally, prune early rather than late. Pinch or clip buds and small shoots today rather than waiting to saw off branches tomorrow. Someone once asked an experienced gardener how he kept everything so neat and trim. With a smile, the old gentleman merely held up his stained and calloused thumb and forefinger, evidence of years of careful pinching. No verbal testimony was necessary.

THE GOAL: A NATURAL LOOK

Painting a wall with plants

"I have a predilection for painting that lends joyousness to a wall," wrote French impressionist painter Pierre Auguste Renoir. A joyous wall is also an excellent reason for training plants flat against a vertical surface. But the use of espaliered or flat-trained plants is not limited solely to giving pleasing shape, color and texture to an otherwise featureless façade. The plants also serve practical functions. They frame windows and doors, disguise architectural oddities, increase privacy along a fence, and extend garden space upward in cramped quarters such as the narrow area between a house foundation and a walkway. And, through the use of clever *trompe l'oeil* devices, the wall planting on pages 30-31 even manages to add the eye-fooling image of a long arbor to a garden that otherwise does not have a view.

One of the first decisions to make when planning a wall of plants is whether to prune them in a formal manner with perfect symmetry, as in the candelabra at right, or to opt for an informal look in which plants, though two-dimensional, appear to follow their natural inclinations, as on pages 26-27. For gardeners whose artistic preferences run to nonobjective painting, a third possibility is an abstract design such as those on pages 28-29. No matter which style you choose, it is important to remember that great gardens, like great paintings, are not made in a day: several years of work may be required to complete your masterpiece.

As your pruning progresses, you will find that nature challenges your skill as an artist as well as a craftsman. Occasionally you will need to cast a critical eye over your plants and ask yourself whether they have grown too bulky for their space or whether your well-wrought design has, with time, turned into a shapeless tangle. And even if you intend to cover an entire wall with foliage, you will still need to keep windows free and gutters open. As renowned California landscape designer Thomas Church put it, you must continually employ your pruning shears as a brush, adding or eliminating strokes "which busy nature adds without permission."

A single six-year-old pyracantha shrub trained along a wooden trellis embellishes a flat expanse of whitewashed wall with a classic seven-branched candelabrum.

Architectural elaborations

Espaliered plants are ideal devices for masking the physical flaws of a house or heightening its best features. They can be used to frame a pretty window or to draw attention away from an awkward one. They can break up the line of an overly long wall or extend one that seems too short. The boxy wall of the house below, for example, is reproportioned by two flat-trained cherry trees that soften and hide the corners of the building.

The branches and foliage of two 20-year-old Montmorency cherry trees trace patterns on the surface of a stucco wall while turning a window into the centerpiece of an espaliered bower. One branch is trained above the window—a process that took five years—to lead the eye across and break the wall's 10-foot height.

A flowering quince bush, one of three espaliered against the west wall of a Cape Cod cottage, focuses attention on a shuttered, curtained window. The 8-foot-high shrub, normally grown as a spreading specimen, has been carefully pruned to resemble a vine and gives the wall a richly textured look.

A white-blossoming Chinese wisteria vine rounds the corner from the backyard of a townhouse and runs in a horizontal line under the upstairs windows, bringing a sense of unity to the random fenestration. At one point it forms a T with a narrowly pruned boxwood column, which appears to anchor the vine to the ground.

Natural inclinations

Though severely pruned to grow in a two-dimensional plane, espaliered plants need not necessarily have artificial shapes. Careful selection and placement can impart a natural, three-dimensional look. To achieve this effect, choose plants that have foliage all along their branches plus a distinctive structure that will be easy to maintain if branches must be shortened. Also helpful are graceful contours, like those of the Japanese holly at right.

A magnificent 40-year-old Japanese holly, pruned to accentuate its leftward lean and reveal its venerable limbs, gives the wall of a house the drama of a windswept mountainside. Found abandoned at a construction site, the holly conformed so quickly to its two-dimensional plane that its training wires were soon removed.

Seen in the early spring before its leaves are fully out, a 10-year-old espaliered dogwood reveals the natural fan-shaped structure of its branches. Though trained to grow flat against the wall, the tree appears to be so normal that it seems fully rounded. Espaliered dogwood produces relatively few branches.

Abstract expressions

Because espalier is essentially an artistic method of pruning, allowing the gardener to construct new forms from living plants, it is not surprising that some practitioners try their hands at intensely personal designs. Often easier to imagine than to execute, such designs need constant care. It took a year and a half of cutting and pinching to turn a free-running Carolina jasmine vine into the topiary-like pompon pattern below.

Heavily pruned into a design of dense spheres that resemble herbaceous balloons, an evergreen Carolina jasmine rounds the corner of a California fence. The two-year-old espalier must be trimmed every few weeks through spring and summer or it will run off in all directions. Garden ties hold the vine to staples.

Cutting diagonally across a wall, a climbing hydrangea is intersected at 1½-foot intervals by parallel branches trained along guide wires. The calligraphic design must be pruned every six weeks to eliminate unwanted shoots. If left unpruned, the 15-year-old vine would be shorter and bushier; pruning forces vertical growth.

A man-made vista

An "illusion trellis" is the name given by the gardener who created this example of the art of visual deception called trompe l'oeil. An espaliered yew forms the illusory pathway through an arbor that appears to recede but is actually flush against the wall. Two more espaliered yews flank the sides. Completing the picture is a miniature urn, scaled down from the urns in front.

The right technique for the job 2

"Plants are like children," said the nurseryman. "If you shape them right when they are young, there's less correcting to do when they are grown. The important thing is to shape them judiciously." The nurseryman knows that each plant in the world is unique, with its own idiosyncracies, but that it shares with other plants of its kind certain ideal norms of growth. He prunes to encourage one characteristic without discouraging another, adjusting basic rules to fit each situation. This is partly a matter of understanding how plants grow and how they respond to pruning *(Chapter 1)*. Then it becomes a matter of selecting the right technique and tools for the job.

The basic pruning tool is a pair of pruning shears, which comes in two designs, each with its own devotees. In one, the blade-and-anvil type, a straight, sharp blade presses against a flat bed of soft metal; in the other, a hook-and-curved-blade type, the two parts slide past each other like the blades of a pair of scissors. The latter tool is heavier and more expensive if you buy a good one, but it makes a closer, cleaner cut and therefore reduces the danger of damage to the branch or stem. Also, its pointed blades make it easier to fit into tight places. But the blade-and-anvil type, it should be noted, requires less strength to operate. For special kinds of pruning there are two additional kinds of hand shears. One has a ratchet action requiring little pressure and is excellent for light work, like pruning roses. The other is a pair of cut-and-hold clippers that grips the stem as you cut, intended primarily for gathering flowers.

Good pruning shears will work well on stems and branches up to about half an inch in diameter. For larger sizes you will probably need a pair of lopping shears, which is a long-handled pruner. Usually these are simply larger, heavier versions of the hook-and-blade or anvil-and-blade shears, but sometimes they have gears or levers to increase the amount of force you can bring to bear.

For cutting branches too big for ordinary loppers, you will need

A bountiful crop of ripe apricots is edible evidence of careful pruning. Earlier, while the apricots were green, the grower thinned them out to increase the size of the fruit and lighten the weight on the branch.

CLIPPING SMALL BRANCHES

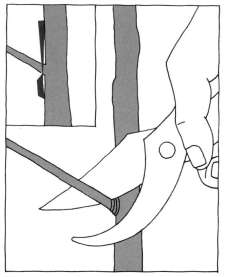

To prune a shoot or branch up to ½ inch in diameter, insert it as deeply as possible into the open jaws of scissors-type pruning shears. Place the thin, large cutting blade flush against the main stem from which the branch is to be cut away (inset). Then in one clean stroke, cut through. Do not twist the shears; they may bind and produce a ragged cut, and you may throw the blades out of alignment. If necessary, smooth the edges of the cut with a sharp knife.

a pruning saw—a tool that differs from a carpenter's saw. Its teeth are angled widely to keep them from binding when cutting through sappy green wood. In addition, a pruning saw generally cuts on the pulling stroke rather than the pushing one—an advantage when the branch you are cutting is above you. Pruning saws come in a number of styles. One type has a curved blade for reaching into confined spaces. Another has a blade that can be set at various angles, like the blade of a jigsaw, making it possible to cut close to a branch or trunk in hard-to-reach places. A larger pruning saw, favored by tree surgeons, has a blade with teeth along both edges—one edge for fine work, the other for coarse. But amateurs using this saw tend inadvertently to gash the bark of nearby branches with the unused teeth. Unless you contemplate extensive work, a small pruning saw with a single set of teeth is probably the best choice.

These three tools—pruning shears, lopping shears and a pruning saw—will see you through most pruning chores. But three other tools should be mentioned. One is a pole pruner or pole saw, essentially a pole with a cutting blade and saw attached; the blade is operated by a pull cord. A pole pruner and saw is especially useful for removing the top branches of fruit trees; the fruit is easier to harvest if top branches are kept within reach. A second possible addition to your pruning kit is a pruning knife with a hook at the end of the blade. Professionals use these to remove small twigs and to smooth and trim the edge of pruning cuts. For most purposes a pocketknife will do as well. Finally, for topiary work and for shaping ornamental hedges, you may want a pair of long-bladed hedge shears. They are undeniably efficient for creating smooth, flat expanses of greenery, but they should not be used for any other kind of pruning. They do not cut precisely and should only be used on plants so well supplied with vegetative buds that they can withstand brutal treatment. For large hedges, electric hedge shears are worth the expense. But it is the only power tool you should use in shaping plants. Do not under any circumstances use a power saw to remove branches; it is dangerous, especially if you are working in a high place.

All of these tools should be of the best quality you can afford and should be kept in top-notch condition. Clean them, oil them and sharpen them. If you have used them on diseased wood, disinfect them with denatured alcohol, a solution of household bleach or the flame of a butane cigarette lighter or a gas range. Keep the blades of shears and loppers adjusted and tightened, because blades out of alignment will crush stems and branches instead of cutting them—and a bruised branch is an invitation to trouble. Bent blades are hard to straighten, so never cut with a twisting motion; the torque combined with the pressure is almost sure to damage the blades.

So much for the tools; now for the techniques. In all pruning techniques there are precautions to take for the health of the plant. First and foremost is this: avoid stubs. When you cut back a branch, cut just above a bud. If you make the cut at a 45° angle in the same direction as the bud's growth, you can cut close with less chance of damaging the bud that you want to let grow. If you are thinning by removing an entire branch, cut as close to the parent branch as possible. With scissors-type pruning shears, this means positioning the shears so that its wide cutting blade rests against the parent branch, producing a cut almost flush with that branch's surface.

If you are using a saw to remove a larger branch from a tree, do the job in stages so the weight of the falling branch will not split the trunk or peel away its bark. First, if the branch is heavy with smaller foliage-laden branches, get rid of as many of them as you can—along with the end of the branch itself. Then, with a saw, make a shallow undercut about 6 inches out from the trunk, taking care not to cut so deeply that the saw binds as the branch sags. After that, saw off the branch from the top and slightly out from this undercut. As the limb falls, the undercut will keep the bark from peeling back. Finally, remove the stub, keeping in mind the tree surgeon's adage: If you can hang your hat on a stub, it needs further trimming.

Having made the cut, you may want to follow it with some post-operative care. As in all surgery, a clean wound heals quickest. You can speed the healing by paring off ragged edges with a pruning knife or pocketknife. After that, in times past, the wound was usually painted with a tree-wound dressing. The question of whether or not to paint is one on which experts disagree. Experiments suggest that tree paint does little good and may actually do harm by trapping fungus spores beneath the paint's gummy surface; one large arboretum near New York City has given up the practice entirely. And many tree surgeons paint wounds only on the customer's insistence, for cosmetic reasons. "If a customer thinks wounds should be painted black, we paint them black," said one veteran of the business. Painting, it follows, can be ignored in home pruning, so long as you are ready to treat the wound with fungicide if trouble develops.

Most people think of trees when they think of pruning—and with good reason. Pruning a neglected tree can be a major undertaking. But if you prune a tree properly when it is young, you will greatly reduce the amount of maintenance pruning it will need later. Especially important is the pruning of a young deciduous tree when you plant it, for pruning gets the tree off to a good start.

This first pruning varies with the form in which the tree is sold:

LOPPING THICK BRANCHES

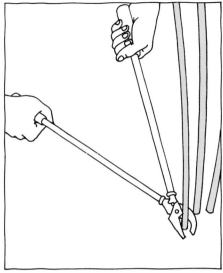

Lopping shears make quick work of canes and woody branches up to 1 or 1 ¼ inches in diameter. Their long handles increase leverage and deliver more cutting power to the blade than is possible with hand shears. The length of the handles also makes it easier to reach into the center of a large, dense shrub. To make cutting easier, you may want to buy lopping shears that are equipped with gears or levers that multiply the pressure you can bring to bear on a cut.

in a container, balled and burlaped, or bare-rooted. Often a tree grown in a container needs no pruning at all; its roots should be intact—you can verify this when you remove it from its pot. Balled-and-burlaped trees that have been root-pruned with several transplantings generally need only light pruning. Any roots that are broken or exposed should be cut off and the top should be reduced proportionately. Usually it is sufficient to cut back each branch to the first bud or node in from the tip. But if the tree has never been transplanted and has been dug up for you, it will have suffered severe root damage and must be pruned accordingly. To restore the balance between the remaining roots and the top, cut back the top by about two thirds, either by removing all but a third of its branches (exempting, of course, the central branch, or leader) or by removing two thirds of each individual branch.

A SPEEDY RECOVERY

Trees sold bare-rooted require the same Draconian treatment. This applies even to bare-rooted fruit trees, which are sometimes sold as one-year-old "whips," with no branches at all. When cut back two thirds, a fruit-tree whip may be reduced to nothing more than a few healthy buds—not an impressive sight. But take heart. In a few months the tree will regain its former size and be flourishing. (Slower-growing evergreens are almost always sold in containers and rarely need any pruning at planting time.)

After a year or so a second critical pruning occurs, establishing the ultimate shape of a shade or fruit tree. Depending upon its intended use, this shape may be essentially spreading or upright. Fruit trees, as well as many ornamental trees and those that will

REMOVING LARGE BRANCHES

1. *Canes and branches more than 1½ inches in diameter call for a special technique and a pruning saw. To sever a large branch without tearing the bark, make a shallow cut on the underside 6 inches or so out from the trunk. Saw through the branch from above, just beyond this cut, leaving a short stub.*

2. *Examine the stub to locate the narrow band separating the different bark patterns of branch and trunk. Make a shallow upward cut through this band. Then cut off the stub from above, keeping the saw blade within the band. Trim the cut with a knife.*

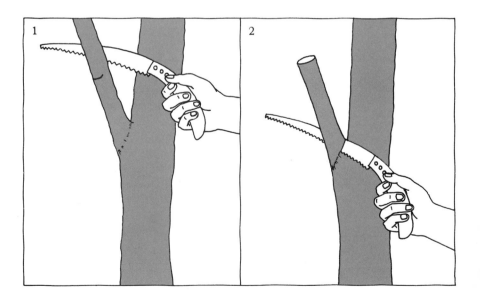

serve as screens or windbreaks, are pruned to begin branching not far above the ground. Their main lateral branches, the so-called "scaffold" limbs, are chosen accordingly. This selection begins when the tree is about a year old, when all but three or four of the shoots that will develop into lateral branches are removed. The chosen shoots should be spaced well apart and come out from the central leader, the eventual trunk, at a wide angle. If these potential scaffold branches seem spindly at first, prune them back to strengthen them and remove some of their side shoots so they have less weight to support. Finally, if you want the tree to be more compact, prune back the central leader to stimulate lateral growth.

Shade trees are handled somewhat differently. Because you will want to walk or plant under them, they are pruned to provide headroom. The first scaffold branches begin at a height of 6 to 8 feet. When the tree is tall enough to produce branches at this height—in three or four years—select three or four promising laterals and cut off their tips at leaf nodes. This will encourage them to grow stronger and thicker. Do not, however, remove unwanted lower branches yet; their leaves are still needed to supply nourishment to the chosen scaffold branches. These lower branches may die back of their own accord, but you can help nature by removing them in a year or two, as the tree gains further height and foliage.

Some of the cuts made to establish a tree's framework are identical to those done later, and routinely, to maintain the health of a full-grown tree. In the course of creating scaffold limbs you cut out any vertical growth that competes with the main leader, or any branches that rub against each other. You also cut off any branches whose angles with the trunk are very narrow—a narrow angle makes a weak crotch that is vulnerable to splitting. All these are part of the regular maintenance pruning that keeps a tree well shaped and healthy. You can do this periodic trimming yourself, although the removal of branches beyond the reach of a pole pruner is more safely handled by a professional. The trimming can be done at any time of year, though if flowers or fruit are involved you should prune immediately after bloom or harvest. At many arboretums, maintenance pruning is done in winter, when other chores are light. But it is easier to spot dead or diseased branches in summer. Generally, you do maintenance pruning whenever you note a problem.

Maintenance pruning can be done in three stages. First, and most urgent, cut away any dead, damaged or diseased wood. Second, do preventive pruning—the pruning that forestalls future trouble. Under this heading comes the removal of crossed branches, weak crotches and competing leaders. Also get rid of drooping

CUTTING HIGH BRANCHES

For branches high overhead, the safest tool is a pole pruner. It allows a gardener to reach as high as 15 feet without resorting to a ladder. One type combines a detachable pruning saw, for cutting larger branches, with a pruning hook and rope- or rod-operated blade for branches under 1 inch in diameter.

THREE-STEP MAINTENANCE

branches, which bear few flowers and poor fruit, and cut back or remove branches that threaten to interfere with utility lines or rub on the house. Be on the lookout for two kinds of extraneous growth that afflict deciduous trees—suckers and water sprouts. Suckers are swift-growing shoots that appear around the tree's base; water sprouts, which resemble suckers, grow straight up from the lateral branches. Both, on occasion, can be useful. A sucker in time can replace a diseased or damaged trunk, and a water sprout can fill a hole left by a fallen limb. But ordinarily both of these random growths should be removed as soon as possible. Water sprouts tend to spoil the look of a tree; they and the suckers rob it of nutrients. If the tree has been grafted, suckers may be coming from the wild rootstock. Being stronger, the stock could become dominant and thus defeat the whole purpose of the graft *(Chapter 3)*.

The third and final stage in maintenance pruning is thinning to open up the interior of the tree and prevent its foliage from becoming too dense. Often the work done in the two previous stages will eliminate the need for the third. But if thinning is necessary because a tree has been long neglected, it is accomplished by removing the less desirable branches entirely.

Maintenance pruning sounds arduous, but if done regularly it is not likely to take more than an hour or so per tree during the course

THE KINDEST CUT FOR PRUNING

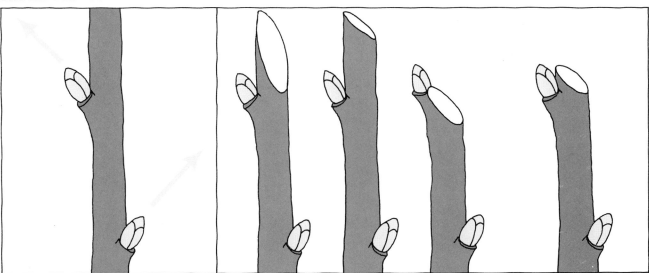

To prune properly, first select a bud pointing in the direction you wish new growth to go in. Using clean, sharp pruning shears, cut just above the bud on a slanting line parallel to the bud's direction of growth.

The three pruning cuts on the left are incorrect; the one at the right is made properly. The cut at the far left slants too steeply, exposing too much vulnerable heartwood. The next cut is too far from the bud; the

stub will not heal. The third cut comes too close to the bud and may cause it to dry. The ideal cut, right, starts slightly above the tip of the bud and slants down at a 45° angle, ending level with the base of the bud.

of an entire year. There are, however, exceptions. Fruit trees, for instance, take more care because a good fruit crop depends on consistent and precise annual pruning.

Fruit trees have traditionally been trained to branch out quite close to the ground so their fruit will be accessible for spraying and harvesting. While various systems can be used to achieve this goal, one called the central-leader system is particularly good for home gardens. It produces trees that are strong but not so tall that harvesting fruit becomes hazardous.

To start a central-leader fruit tree, prune back the whip at planting time to a height of 3 or 4 feet. This will cause a large number of buds near the top to start growing. The following year, in early spring, select a central leader and three or four laterals for the scaffold branches, making sure the laterals come out from the leader at a wide angle and at different heights. Remove any branches that might compete with the leader or laterals, but keep as many small noncompetitive branches as you can. Their leaves will help the tree set flower buds—and an early flowering tree has a double advantage. It produces fruit earlier, and because its energy goes into fruit production, the tree itself remains smaller and easier to manage.

The following spring, select two or three additional lateral branches above those already chosen, and prune back the tips of the central leader and the laterals to encourage side shoots. From this point on the only shaping the tree is likely to need will be control of the top growth. In the past, some experts recommended cutting off the central leader above the last main lateral as soon as the tree's scaffold was established. But most now prefer a more natural treatment; in a garden, the tree looks more treelike if it has a central leader. To limit the height of the tree, a strong central leader can be replaced with a weaker upright shoot. This will look like a central leader but will not reach the previous height for several years. Then the entire process can be repeated.

For the next few years, until the tree begins to bear fruit, prune it sparingly, for heavy cutting delays bearing. Apple trees generally begin to bear in their third to fifth year, pear trees in the fourth or fifth year and peach trees in the third. After the tree has begun to bear, you will once again begin to prune, but for a different purpose: to keep the tree yielding steadily. This pruning is dictated by the way the fruit is produced. Apples, pears, cherries and apricots grow on spurs—short stubby offshoots of the lateral branches—that bear for many years. Until they begin to bear, the spurs are hard to distinguish from ordinary offshoots. But at bearing age they thicken, and creaselike rings encircle them at intervals, marking the yearly

TREATING A TREE WOUND

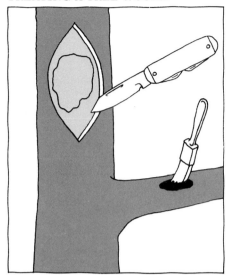

To help a tree wound heal, cut away the ragged edges of the bark with a sharp, clean knife. Shape the wound into a smooth oval, pointed at both ends. This helps the bark shed water that might cause rot and also speeds the growth of healing callous tissue by allowing the tree's vertical conducting channels to function normally. If the wound is on the upper surface of a horizontal branch, coat the wound with a tree paint formulated to prevent the growth of water sprouts.

What to prune and why

Any unnatural or unwanted growth of a branch, shoot or root is a candidate for pruning; allowed to develop, it may threaten a plant's health and beauty. Obviously it is better to prune early, when the problem first appears, for a young branch is easier to cut and leaves a smaller wound that heals more rapidly. It also leaves a smaller scar—an important consideration with ornamental trees and shrubs.

Early recognition of a problem growth can sometimes eliminate the need for pruning entirely. A branch with a weak crotch, for example, can be trained to grow at a wider angle by bracing the branch while it is still young and supple. A fruit tree can be prevented from developing permanently sagging limbs by not permitting too much fruit to weigh down young branches.

Other problem growths can be avoided simply by following good gardening practices. For example, the roots of a tree are less likely to girdle the trunk if they are carefully spread outward and downward when the tree is planted.

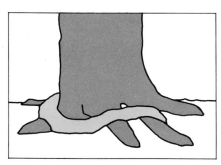

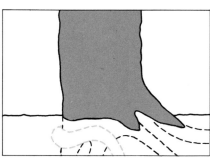

A root that girdles a trunk will eventually strangle the tree by cutting off its life-supporting flow of water and nutrients. Sometimes visible aboveground (top), potentially girdling roots can also be located beneath the surface of the soil. The telltale sign is that the trunk rises without a natural flare (bottom).

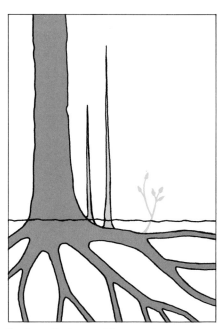

Suckers that sprout from roots near the base of a plant grow faster than the plant itself. They rob it of water and nutrients and may eventually displace it—a particular problem if the top of the plant is grafted on. Occasionally suckers are left to grow if they are needed to replace old and unproductive stems of flowering or fruiting shrubs.

Water sprouts—suckers that sprout from a tree branch or trunk—are so soft that they provide an entry point for aphids and other pests. Though fast-growing, they seldom produce any flowers or fruit and should be removed except when one is needed to replace a branch that has been lost by storm damage or disease.

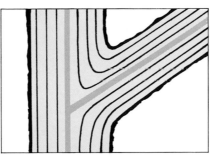

A branch that grows straight up or down, contrary to a tree's natural form, can upset more than its symmetry. The unruly branch can crowd other branches or deprive them of sunlight. On fruit trees especially, the vertical branches should be removed, because a branch that grows straight up will not produce fruit.

The narrower the crotch, or angle, between a trunk and branch, the greater the risk of the branch splitting from the tree. A wide crotch is strengthened by a broad band of connecting layers of wood (top). In a weak crotch, the trunk and branch press together, sandwiching a layer of bark between them (bottom).

When many branches rise from the trunk at the same level, the tree will be very crowded after they have fully developed. Anticipate how large a tree will be 10 or 15 years after it is planted, and leave only those lateral branches that will have space to develop properly. Remove the branches with the narrowest crotches.

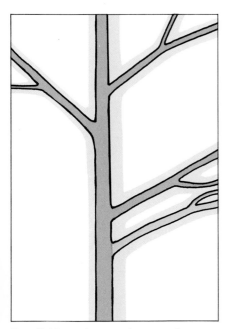

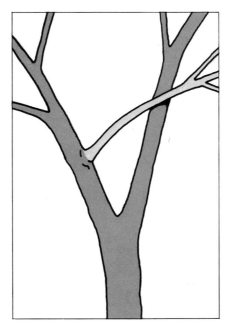

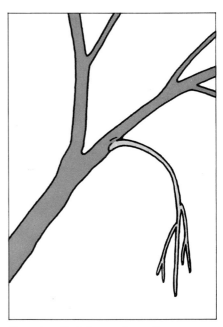

Parallel branches too close together can be injured by blowing against each other on a windy day. On fruit trees, the fruit will pull the branches down, and shade from the upper branch will prevent the full development of fruit on the lower one. The danger in parallel branches is less apparent when a tree is young.

A branch that crosses over another branch will rub against it repeatedly. Wounds that result invite disease, for the constant rubbing does not give them a chance to heal. A crossover branch will also prevent sunlight from reaching the center of the tree and, on fruit trees, it will bump against and damage developing buds and fruit.

Branches that droop when other branches of a plant extend upward are weak and slow growing. A drooping branch is often caused by fruit growing too near the tip of a young branch. It will bend under the weight of the fruit, and if the branch is allowed to remain bent, it will never recover its upright position.

growth. The object in pruning these trees is to keep them vigorous so new spurs will continue to form, to keep them open so light reaches the fruit to ripen and sweeten it and to make it easier to spray. On such trees, branches that begin to droop from the fruit's weight should be cut back so they continue to produce.

Spur-bearing trees can also be made more productive by two additional kinds of pruning. When a spur ceases to bear it can often be rejuvenated by removing half of it, cutting through it at one of its thickened rings. A new shoot will form at the cut and become a new spur. And trees that bear every other year—as some apple trees do—can often be encouraged to bear annually by removing some of the fruit on each spur. The remaining apples will be larger, so you will not lose part of your crop. More important is the fact that the spur will produce flower buds that in turn will yield fruit the following year.

Peach trees and nectarines produce fruit in an entirely different way. Instead of growing on the same spurs year after year, the fruit forms on shoots of new growth, usually in the shoots' second year. But having borne once, these shoots never bear again. It is essential therefore to remove them, not only because they clutter the tree but because they use space and nutrients needed for the development of new fruit-bearing shoots. In this kind of pruning, called simplification, a group of side branches extending from the scaffold branches are cut back to three to five shoots. The remaining shoots will bear fruit and the buds on these shoots will produce new shoots, ensuring fruit the following year.

INCREASING THE CROP

REJUVENATING FRUIT SPURS

The fruit-bearing spurs of apple and pear trees normally produce fruit for about 10 years and then become barren, but pruning (red bars) rejuvenates them. To prune an apple spur (near right), cut it back to one of the thickened rings that mark a year's growth; these rings generally occur at ½-inch intervals. This will force new shoots and buds to form on the remaining part of the old spur. On a pear spur (far right), whose growth is marked by a gradual knoblike thickening at the end of the spur, cut through the knob, encouraging new buds to form at other points on the surface.

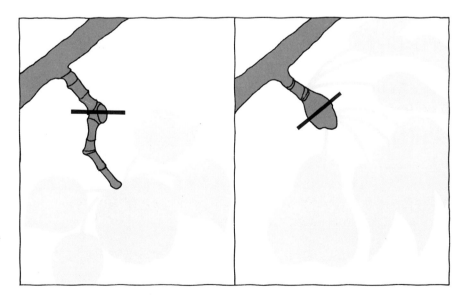

Peach trees, in addition, are sometimes thinned to remove part of their crop—they tend to bear more fruit than can properly develop. To get better fruit, thin them so that only one peach remains for every 6 to 10 inches of the shoot. But do not perform this operation until after the "June drop," when fruit that has pollinated poorly falls off of its own accord—at which point the real size of the crop can be judged.

Even with thinning, fruit trees will occasionally produce so heavily that the fruit interferes with optimum growth. For long-term vigor, apple and pear trees should grow 10 to 12 inches a year, peaches, 12 to 18 inches, until the mature height is reached. If they do not grow this rapidly, they can be stimulated with heavier pruning or fertilization.

Citrus trees are evergreens and in general need less pruning than their deciduous counterparts. Orange and lemon trees develop attractive shapes without meticulous training and can actually be shaped as you would a hedge, by shearing off growth you do not want. While they are young you can pinch off growing tips to keep them compact. Ideally, for an easy harvest, the trees should not be allowed to grow more than 9 or 10 feet high. Lemon trees especially grow rapidly in a mild climate and need to be kept pruned within bounds; they are also apt to develop suckers and water sprouts that should be removed. You may also want to thin out some of the lateral branches from time to time, to make it easier to reach internal fruit. And you should, of course, prune out dead wood.

Although fruit trees are best when young, pruning can sometimes revive an old and neglected one. This is done in several steps. The first year, cut out dead and broken branches and water sprouts; then remove a few of the heavier branches to let more light into the tree's center. The next year, prune back all the branches moderately, and feed in early spring with a balanced fertilizer such as 10-10-10. In another year or so, the tree should be bearing again.

Shrubs, both deciduous and evergreen, are far easier to prune than trees but present their own set of problems. They require regular pruning to keep them from getting out of hand. The main pruning chores are chiefly concerned with scale. An all-too-familiar sight, especially around older homes, is the overgrown plant that blocks the first-floor view. Many owners are afraid to cut back these plantings, lest they kill them. But in fact, most shrubs, especially deciduous ones, thrive under heavy pruning, and the flowering kinds may actually flower more plentifully.

Deciduous shrubs differ from their evergreen counterparts in one important respect. They continually send up new canes or stems

THINNING FOR BETTER FRUIT

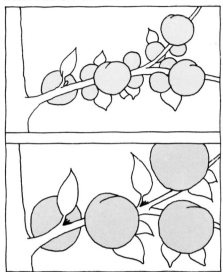

Some of the green fruit that crowds the branches of a fruit tree should be thinned out after the trees have completed their natural fruit "drop," usually in June (top). Thinning prevents branches from sagging or breaking, and the fruit left to mature is bigger and tastes better (bottom). On some trees that normally fruit only in two-year cycles, thinning also encourages an annual crop. Reducing the fruit reduces the number of fruit seeds, which contain a hormone that inhibits flower buds.

from their bases and should be encouraged to do so—every stem or branch can eventually be removed to make way for a new one. This process of renewal pruning begins when the plant reaches maturity. Until that time the only pruning the shrub is likely to need is the occasional pinching of terminal ends to encourage branching.

The schedule for renewal pruning is dictated by the shrub's time of bloom. Those that blossom in the spring are pruned right after the flowers fade, so the plants' energy is channeled immediately into the production of new growth and flower buds for the following year. With summer-flowering plants, it is usually better to wait until the following spring, when new growth starts, to remove old stems. Fall pruning can be hazardous; it sometimes stimulates new growth at the very time of year when you do not want it—just before winter frosts arrive to kill the tender shoots. In addition, if you prune out old stems in the fall you may lose some winter color, like the berries on barberry and viburnum.

Renewal pruning is also governed by a deciduous shrub's habit of growth. Normally, as new growth appears, old growth dies back. But on fast-growing plants like forsythias and some of the spireas, there is so much overlap in this cycle that the plants quickly look dense and cluttered. For the sake of a shrub's health as well as its appearance, help nature along by removing old growth periodically. You can cut a shrub back to the ground, but usually, for appearance's sake, only one third is removed each year. In this way you can totally renew an aging plant in three years.

On most mature shrubs you can spot older stems easily. Their bark is dull and flaky, while the bark of young stems is smooth and bright. But a special problem is presented by lilacs—which fortunately do not usually need heavy pruning. In fact, if you prune back lilacs before they bloom you are sacrificing flowers, which are borne on branch tips. The old and young stems of lilacs look almost alike. You can, however, spot older stems during the plant's period of bloom and tag them for later removal—the flowers are fewer and less vigorous on older stems.

SHAPING UP A SHRUB Besides renewal pruning, a shrub may need reshaping from time to time. If it seems too sparse and rangy, you can tighten it up by pruning back its stems to inward-facing buds, thus forcing new shoots to grow in that direction. Conversely, by pruning back to outward-facing buds, you can enlarge a compact shrub. You may also need to alter the shape of a shrub that has a tendency to produce dense, twiggy growth at its tips. This growth should be thinned out, so that light and air can reach the plant's interior.

The same general rules for pruning deciduous shrubs apply also to flowering evergreen shrubs. The best time to prune is after flowering, and the frequency depends on each plant's habit of growth. Fast-growing evergreen shrubs like the ligustrums grown in mild climates need regular pruning and thinning to prevent them from taking over your whole garden; in fact, they will stand shearing, though selective pruning gives better results. At least once every three years, cut out older stems and prune back the remaining ones to outward-facing buds. Slower-growing broad-leaved evergreens like rhododendrons, mountain laurels, camellias and azaleas need less control. Some experts say rhododendrons should not be pruned except to remove spent flower heads. These should indeed be removed to encourage better flowers the next year—and they

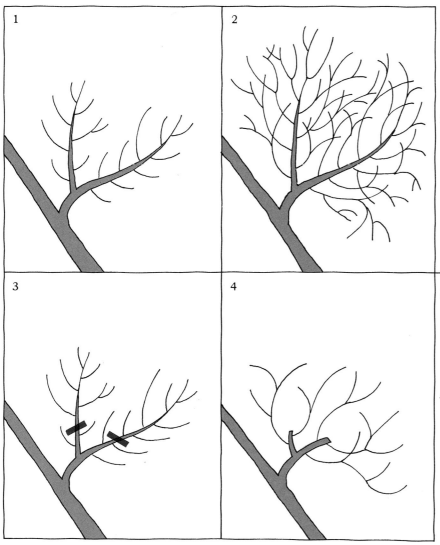

SIMPLIFYING SIDE BRANCHES

1. *On many plants the natural pattern of growth produces a complex structure of forking side branches. From these branches grow the shoots or spurs that produce a plant's flowers and fruits.*

2. *If side branches are not pruned, the plant resembles a mass of tangled antlers. Its crowded interior branches receive inadequate light and air and may die, while its crop of flowers or fruits, which are produced at the tips of the shoots, will be disappointingly small.*

3. *To simplify a branch and make room for healthier growth, cut back the tips of forking side branches so no more than three to five shoots remain near the base of the fork. Leave no stubs, so the plant will have a natural, unpruned look.*

4. *With ample light and space, the simplified branch develops new, sturdier side shoots. These in turn produce buds that flower more profusely and yield larger, handsomer crops of fruit. The process should be repeated yearly.*

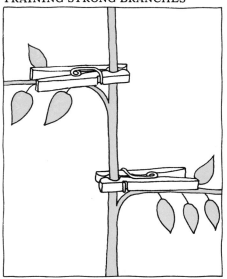

Use spring-clip clothespins to train the tender side shoots of young trees to grow horizontally and widen the crotch. In late spring, press the shoot down into position and clip the clothespin to the whip, or main stem, just above the shoot. Remove the clothespin later, in midsummer, when the shoot has matured and hardened enough to maintain the horizontal position. A wide-angled branch is strong and on fruit trees will bear at an earlier age.

should be removed with care, snapped off with your fingers rather than cut, so as not to damage the vegetative buds just beneath.

But rhododendrons do get leggy, and heavy pruning does not harm them. In England during World War II, a number of 100-year-old rhododendrons, many with trunks a foot in diameter, suffered bomb damage and had to be cut down. Most were soon bushy plants again. How much you can safely cut back a rhododendron depends on its vigor. If you are in doubt about its vitality, prune it gently at first, cutting back a few branches about a third of their length each year. But if it is growing well, you can remove a third to a half of the old stems each year. This is also true of azaleas, although they are more likely to stay within bounds with nothing more than light pruning and pinching. Camellias and mountain laurels may be shaped as necessary after flowering.

A third kind of evergreen shrub, the needled conifers that so many homeowners find ideal for foundation plantings, also grow so slowly that they seldom need to be curbed. Pines, spruces and firs should not in fact be cut back at all. If they outgrow their sites they must be moved or cut down. Many others, however—the hemlocks, yews, junipers and arborvitae—can be reshaped as much as is desirable. Many homeowners trim them formally into globes and spires with hedge shears, but they look more natural if they are simply pruned back selectively to nodes or branching points. In any case, do not cut back beyond needle-bearing side branches or the branch will die. And remember that any branch that is removed completely is gone forever; it will not be replaced.

Probably the most pruned of all shrubs are the roses. Chances are, if you see a rose gardener without his pruning shears, it is only because his hands are full of spraying equipment. The pruning varies with the type and variety of rose, and some of the pruning chores are quite demanding. But none is mysterious. Roses, like all shrubs, benefit from being kept open and uncluttered, and from being constantly renewed.

Pruning time for most roses is early spring, just as the buds begin to swell, and the first task is to cut away any winter damage—broken, dead or discolored branches. Prune back to healthy buds, preferably facing outward so the center of the plant will be more accessible for subsequent spraying. This will stimulate the production of side shoots, and it is here that rose pruning becomes an art. Roses, like peaches, bloom on new growth so you want to encourage as much new growth as possible. At the same time, you do not want the rose to produce more blossoms than it can readily support. A pencil-sized cane will support about two buds, or two roses; a thumb-sized cane up to 10 or 12 buds. In pruning back to achieve this balance you will strengthen and thicken the stems.

For rose growers, the whole point of roses is the blossoms, and much pruning is designed to achieve more, or more spectacular, blooms. If you want large blossoms, you prune severely; lighter pruning will yield smaller but more numerous flowers. Later in the season, as flower buds develop, you may want to thumb off all but the terminal bud on each stem of a hybrid tea rose. Called disbudding, this process results in a considerably larger flower. In cutting flowers for indoor display, remember that this too is a form of pruning. If you cut carefully you can encourage the plant to blossom more abundantly. On most stems there are two or more five-leaflet leaves. Cut just above one of these, leaving no less than two on a stem. Flower buds should form at the bases of these leaves.

Of all the various kinds of roses, the hybrid teas are pruned most severely. They are swift growing and quickly become leggy, so older canes must be regularly removed. Many of these rose plants are grafted onto stronger rootstocks, so you must learn to distinguish between basal shoots that rise from above a graft and suckers that grow from an understock. The former will provide new canes; the latter should be removed lest they take over. Usually you can identify the graft by the slight swelling at the base of the plant, but if it eludes you, check the leaves. Those of a sucker will be different from the leaves on the grafted plant above.

Other types of roses are pruned lightly in the spring and selectively thereafter. The polyanthas and floribundas, which bear flowers in clusters, should be pruned again when spent flower heads are removed before the plants form rose hips; the canes should be cut back to the nearest buds. The climbers—the large-flowered varieties, the everbloomers and the ramblers—produce their best blooms on canes one to two years old. Consequently, older canes should be removed after they have finished blooming—for the everblooming varieties, at the end of the summer. Tree or standard roses, consisting of small bushes grafted onto tall stems, should be pruned according to the type of the grafted top—if your top is a hybrid tea, for example, it should be pruned severely. But the base of the plant should be pruned only to remove suckers. The venerable shrub roses, generally wild species, are usually left unpruned, although such plants can be renewed by cutting out about a fourth of the oldest canes each year for four years.

Like roses, the fast-growing bramble fruits, such as blackberries and raspberries, are popular in backyard gardens. But in no time at all, if not kept in check, they can become the kind of impenetrable

INSULATED ORCHARDS

In parts of Central Russia, where temperatures can drop to a bitter 50° below zero, dwarf fruit trees are pruned and trained so they will be buried during winter under an insulating blanket of snow. The trees, grown close together in rows, are less than 3 feet tall. They are kept that low by planting them at a 30° angle and by training their branches with wire hooks to stay close to the ground. The Russians report they have more than 10,000 acres of "creeping" apple and sour cherry trees as far north as Omsk, on the 55th parallel. A smaller display of these trees is growing even farther north at the Botanical Garden of the Komarov Botanical Institute in Leningrad.

thicket that surrounds enchanted castles in childhood fairy tales. Almost all of the bramble fruits are perennials with biennial fruit-bearing canes—that is, the canes form one year and bear fruit the next. Spent canes should be removed promptly, particularly if—like the canes of the black raspberry—they may arch and take root.

At planting time and for the first few years, most bramble fruits are trained more or less alike. The newly planted bushes should be pruned back to stubs and then, at the end of the first growing season, a handful of promising canes should be selected and the rest removed. For most fruits, five or six canes are the usual number, but for thornless blackberries the ideal number is three. These canes, which will be 5 to 6 feet high, should be pruned back slightly to encourage the production of fruit-bearing laterals the following spring. The laterals too should be cut back, to a length of about 12 to 16 inches, to concentrate the plant's energy on fruit production. In the fall, when fruit-bearing canes are spent, they are normally removed and new canes are selected for the next year. But there are exceptions. Certain red raspberries bear their fruit in both summer and fall, and bloom and bear on the same canes a second year. On these plants only the top half of each bearing cane is removed.

Grapevines and bramble fruits would seem to have little in common, yet pruning is vital to both of them. Without pruning you will have little fruit from either. The pruning practices of commercial grape growers are an exacting science. But for the home gardener two simpler systems are recommended: the long-cane system and the arbor system, which are both suitable for native

TRAINING BERRY BUSHES

PRUNING FOR MORE ROSES

1. *A bush-type rose such as a hybrid tea rose may be pruned to encourage either more flowers (right) or bigger flowers (opposite). In both cases, in early spring you should first remove any winter-damaged parts of canes as well as canes that cross each other (blue sections). Whenever possible, cut just above buds that face outward. Remove any shoots below the knoblike graft union.*

2. *This much pruning will yield numerous flowers of average size. To keep the bush blooming longer, cut off the flower heads as blooms begin to fade in order to prevent seed formation.*

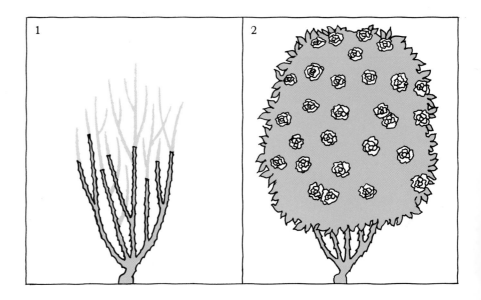

American grapes like the Concords and Fredonias of the Northeast and the muscadine of the South.

In the first two years of training, both systems are identical. At planting time, the vine is cut back, leaving only two buds on the stem. When the shoots from these buds are several inches long, choose the stronger of the two and cut it back also to two buds, removing the other shoot entirely. The following spring, again select the strongest shoot and remove the other. This establishes the permanent trunk. From this trunk laterals will grow and will be trained to expose as much leaf surface as possible to the sun.

If you are using the long-cane system, the laterals will be trained to two wires running between stakes, one wire 3 feet from the ground, the other 5 feet. When the trunk reaches the lower wire, select two lateral canes and tie them to the wire. Then let the stem grow to the upper wire, select two additional canes and tie them to the wire on either side. Remove all the other laterals. Over the next few months keep the trunk pinched back to halt its growth and to channel energy into the canes. The following year cut each of the canes back to five or six buds; these buds will produce shoots, which will in turn produce grapes.

In subsequent years your pruning will have two objectives: to control the size and to maintain the quality of the crop. You limit the number of buds so that the vine does not overbear and produce inferior grapes. And you renew the canes each year so that the vine will continue to be vigorous. Each winter, choose one of the shoots nearest the trunk on each cane as a replacement for the existing

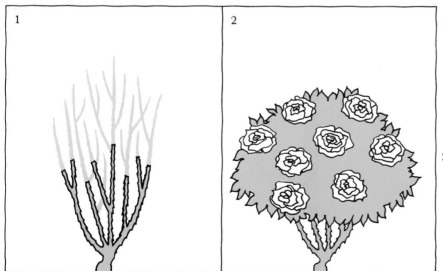

PRUNING FOR LARGER ROSES

1. *In the second method of pruning bush-type roses, you should not only remove injured and crossing canes (opposite) but cut back each cane to half its height (blue sections). Promptly remove any suckers that may develop below the graft union as a result of this more severe pruning.*

2. *The resulting bush will be smaller overall, with stronger canes and fewer but larger flowers. Pinching off flower buds lower down on canes will increase the size of the remaining blossoms still more. For the largest blossoms of all, remove every flower bud except the terminal one on each stem.*

cane, and cut off the old cane beyond that point. The new cane will grow the following year, producing shoots and fruit, and in turn itself be replaced.

When the new canes acquire eight to ten buds, you may want to remove a few of them to limit the size of the crop: two to four clusters per cane is generally an ideal number. Depending on the grape variety, you can disbud near the trunk, near the tip or anywhere in between. Some grape varieties typically produce fruit on shoots close to the trunk, others on shoots six to ten buds out, and still others on every shoot on the cane. Eventually, as the yearly replacement canes begin farther and farther from the trunk, the day will come when the entire lateral will need to be replaced. When you cut it off, new laterals will develop from dormant buds on the trunk. One should be selected for training and the whole process repeated.

GRAPES ON AN ARBOR

The arbor system is an informal version of the long-cane system. The vine is trained up a trellis or arbor, and perhaps over a terrace or balcony to provide shade as well as fruit. Instead of being halted at a height of 4 or 5 feet, the trunk is permitted to grow tall and develop any number of lateral canes. You may even decide to have two or three trunks instead of one. Every year, the canes of these vines should also be renewed and their number of buds limited—but the extent of this pruning will depend on the individual vine. You will have to experiment to discover how much fruit your vine will support. If you want to cover a large arbor for shade, you may not want to regulate the crop by cutting back canes. In that case, you can simply thin the grape clusters.

Compared to grapevines, other vines are child's play to prune. Some homeowners never prune them at all, which is a mistake. An unpruned vine can become quite overgrown and the flowers of a flowering kind will become smaller. In general, vines can be controlled with the same techniques—cutting back and thinning out— used on trees and shrubs. Many of them also benefit from renewal pruning when their stems become thick with age. A diverse group, their time of bloom and habit of growth dictate the time and degree of pruning. Some clematis species, for instance, bloom on new shoots and should be pruned in the spring, just before new growth starts. That notoriously rapid grower, wisteria, can best be kept in bounds by being pruned back severely in the fall—by thinning out side shoots and cutting the remaining ones back to two flower-producing buds. Many gardeners also make at least one extra sweep over wisteria in the summer, cutting off the long stems that wave in the air.

Of all the pruning tasks facing homeowners, probably none looms larger in terms of time spent on the job than trimming a formal hedge. Properly speaking this is not pruning at all, but shearing. It does, however, affect the growth of the plant—and many people do it the wrong way. The purpose of a hedge is to screen from the ground up, which means caring for it so that the bottom remains as healthy and green as the top. This training begins the very moment the hedge is planted. As soon as the plants are in the ground they should be cut back to a height of 4 or 5 inches; the most popular hedge plants—privet and yew—are generally 18 to 24 inches high when purchased. As the hedge grows, it should constantly be pruned back, roughly 6 inches for every 12 inches of growth, until it reaches the desired height. If it is allowed to grow to this height unchecked, the bottom will never have a chance to fill out.

As you cut the hedge back, start to shape it. You may want it to be rounded, or more or less rectangular. But whatever the shape, the sides should slope inward as they rise. The top of the hedge must be narrower than the base so the lower branches will get the light they need. Hedges planted in the shade should, for this reason, slope more than those planted in the sun. The top of the hedge may be flat or rounded; in the north the latter is favored because it sheds snow more readily. Elsewhere the rounded shape may be preferred because a flat hedge, in the words of one Scottish gardener, calls for "a stiddy han' an' a strraight ee." The "strraight ee" can be guided, of course, with two stakes and a taut level string.

After the hedge is the height you want it, it will need several trimmings a year. Fast-growing privet may need as many as three; for others you can probably get by with two. Fortunately, electric trimmers make this task less arduous than it used to be. Some gardeners delay the first trimming until the spring growth has peaked, around mid-June. Others do not wait this long, and in fact the hedge stays more tidy and uniform if you do not let the new shoots get more than 6 inches high. The last trimming of the year should be no later than the end of summer, so the new growth that is stimulated by pruning will have time to harden before frost.

Finally, there is trimming that must sometimes be done, especially with more vigorously growing plants, when the hedge passes its prime. If it becomes dense and overgrown, it must be dealt with mercilessly. The remedy, severe cropping, is often more traumatic for the gardener than for the plant. One gardener faced with a privet hedge that had finally reached its natural limit, between 12 and 14 feet high, took a deep breath and sliced off the entire hedge to a height of 6 inches. In a year or so he had what looked like a brand new hedge, a neat and manageable 16 inches high. Sometimes the more drastic the pruning, the more dramatic the results.

WIDE-BOTTOMED HEDGES

Heads held high on elegant stems

Elegance and style are words frequently applied when gardeners discuss the use of pruning or grafting to create the long-stemmed, large-headed plant forms known as standards. Unlike the artificial shapes achieved by topiary *(pages 90–99),* standards are simply stylized versions of the natural structure of a tree. Properly trained and maintained, a plant grown as a standard can have a normal life span—lantana standards *(opposite)* sometimes last more than three quarters of a century.

In general, small-leaved plants that can be pruned to retain round, bushy crowns make the best standards. However, since the main prerequisite is a stem that is always bamboo-straight and unbranching, a tree such as the Japanese weeping cherry on page 56 also makes a handsome standard. As long as the branches puff out from the top of a bare trunk, the foliage may take any shape.

Training and pruning the supporting stem of a very young plant is not always easy. It may need to be tied to a supporting stake for years to hold it rigidly upright until it becomes strong enough to support the weight of the crown; even more years may pass before the stem reaches the desired height. To speed up the formation of a standard and to increase its chances of survival, many gardeners graft a plant that can form a handsome crown onto the straight stem of a sturdier plant in the same family. For example, a white hybrid tea rose may be grafted atop a vigorous bramble brier root to yield an elegant rose-tree standard.

Whether merely pruned and trained or grafted as well, standards require careful attention to the precarious balance between leaf and root growth that must be maintained in any plant. Since the bare-bones pruning of the stem eliminates foliage that the plant normally exposes to sunlight, you must compensate for the loss by allowing enough top growth at all stages to keep the plant healthy. Remember too that although pruning of the stem will speed its upward growth, the pruning must not be so radical that it reduces the girth necessary to support the top.

A blossoming standard of common lantana thrives 20 years after being planted as a 4-inch cutting. The bushy crown's vigorous growth is spurred by yearly pruning.

Raising your own standards

To grow a standard, choose a woody plant with a straight stem and, if necessary, stake it. Working up from the bottom, pinch off some side shoots—never more than one quarter of the total at one time. Once the stem is tall enough, pinch off the tips of the leader and the three top pairs of shoots. As the shoots branch, remove more growth along the stem and continue pinching back the top. When the crown is fully developed, remove all lower growth.

PONCIRUS TRIFOLIATA
This hardy, 6-foot-tall orange tree was grown from seed.

GERANIUM RED RICARD
This two-year-old supported by a stake is starting to form a crown.

FICUS RETUSA
These fig trees took a year to develop their crowns, now clipped monthly.

RHODODENDRON INDICUM
A stake keeps this azalea's stem straight and helps support its head.

FUCHSIA LORD BYRON
Every winter this tender standard is cut back to one-year-old wood.

55

Grafting to get a lift

To create a standard by grafting, choose plants that not only are closely related but are known to be compatible grafting partners; otherwise the graft may fail. The plants must be joined—in most cases by means of a cleft graft *(page 63)*—so that their cambium layers, underneath the bark, are able to unite. Do not remove shoots and leaves from the lower rootstock until there is sufficient foliage on the upper scion to nourish the plant.

SCION: PRUNUS SUBHIRTELLA PENDULA; STOCK: PRUNUS MAHALEB
This weeping cherry crown was grafted on a cherry seedling.

SCION: ROSA ODORATA; STOCK: ROSA MULTIFLORA
A spring pruning maintains this hybrid tea rose on a 3-foot stem.

SCION: CHAMAECYPARIS OBTUSA GRACILIS; STOCK: CHAMAECYPARIS PISIFERA
A dwarf shrub was used to make the top of this 18-inch-high Lilliputian tree.

SCION: HEDERA HELIX MANDA'S CRESTED; STOCK: FATSHEDERA
Grafting shortened the creation of this pair of ivy trees to a mere three years.

Grafting: one plus one makes one 3

With two deft strokes of his knife, Richard Walter tapered the end of a small shoot of English ivy, checked to see that it was correctly shaped, and gently inserted it into the cleft he had made in the top of a two-foot-high stalk of fatshedera. He squinted at the placement to make sure the English ivy's cambium layer, right under the bark, was lined up with that of the fatshedera. Then he fastened the union with a rubber strip and covered it with a plastic bag to keep the moisture in. "If the two pieces fit together closely and are set just right," he said, "they will knit in a few weeks and we'll have a new plant—one that you can't find in nature."

A retired public park manager, Walter grafts plants for pleasure and occasional profit. "It's so easy to do, I'm surprised more people don't try it," he said. "There's no limit to what you can think up, and it's fun." To prove his point, he selected still another kind of ivy growing in his greenhouse, one with somewhat different leaves. "We'll just stick it up there on top of the fatshedera, next to the English ivy," he said. "That will make an ivy tree with two different kinds of leaves. We will fool a lot of people."

Gardeners and horticulturists have been fooling their friends and benefiting mankind with imaginative—and important—grafting projects for many hundreds of years. The Chinese apparently were grafting plants as early as 1000 B.C. Aristotle mentioned it, and in Romans II, St. Paul used it as a metaphor: "If some of the branches were broken off, and you, a wild olive shoot, were grafted in their place to share the richness of the olive tree, do not boast over the branches." Many of today's techniques were practiced in 16th Century England. Refined and enlarged, they have become, along with the allied craft of bud-grafting, key operations in two huge areas of commercial horticulture: fruit trees and hybrid roses.

Grafting's possibilities are so extensive that it has been the subject of many exaggerated claims. One man boasted of having

A Renaissance gardener on a plaque by Florentine sculptor Luca della Robbia attends to his grafting chores. It is late February, when the sun, crowned here by the sign of Pisces, provides 10½ hours of daylight.

59

grafted a rose onto a black currant to produce black roses, a botanical impossibility. And a New York newspaper once naively and erroneously reported the introduction of a *table d'hote* tree on which, through grafting, such disparate crops as tomatoes, cucumbers, potatoes and apples grew in profusion. It was said to be ideal for the small backyard.

GRAFTING'S FIVE FUNCTIONS

If such absurdities are unhappily beyond reach, grafting's genuine usefulness is not. It may be said to serve five purposes. First, grafting is used to propagate plants that cannot be conveniently or economically reproduced by other means. Most hybrid fruit trees, for instance, do not breed true if grown from seed—they tend to revert to ancestral forms—and cuttings of some varieties will not root. But as grafted plants they perform handsomely, and the same is true of a number of flowering shrubs. Second, grafting makes it possible to combine certain traits like disease resistance, winter hardiness and luxuriant bloom in one plant. New roses are almost invariably grafted onto disease-resistant rootstocks. Third, grafting can alter the appearance or behavior of a plant; many trees are dwarfed this way, while others are made to bloom better. Fourth, it can be used to repair damaged plants and put new life in old stock. Trees in danger of dying because of girdling can be saved by a grafting technique known as bridging. Finally, grafting makes it possible to produce plant forms that could not otherwise exist, like Richard Walter's tripartite ivy.

In essence, grafting involves the wounding of two growths and the arranging of them so they heal together. One of the two growths is called the stock, understock or rootstock. It is the host plant, rooted in the soil and providing nourishment for the other growth, the dependent top section, which is called the scion. In Richard Walter's graft the fatshedera was the stock and the English ivy the scion. Usually the scion is a pencil-slim one-year-old shoot, several inches long, containing two or more buds. But it could be the entire top of a house plant or a single bud attached to a speck of bark. Whatever its size and shape, its function is the same: it provides what will eventually become the top of the man-made plant.

THE INTERLOCKING TISSUE

For stock and scion to unite successfully, the work of the remarkable cambium layer is all-important. This life-sustaining tissue, located just under the bark of most plants *(Chapter 1),* fastens the two parts together. Only plants that have a cambium or its equivalent can be grafted. And no matter what method of grafting is used, the cambium layers of stock and scion must line up precisely. The joining is accomplished by cells that intermingle and, if all goes well, interlock, forming new cambium tissue.

But for this tissue to be created—for the graft to "take"—certain conditions are required. The cambium layers of stock and scion must be joined over as large an area as possible, must be held rigidly together and must be protected from drying out. The operation must also take place at the right time of year. Some kinds of grafts are done in the early spring just before the sap rises in the rootstock, while others are more successful if performed as late as midsummer, when strong growth is under way. For spring grafts, the scion should be dormant. If the graft cannot be performed before the plant starts to grow, the scion should be kept dormant by storing it in the refrigerator. Ideally, for early-spring grafts the stock should be dormant too, although successful grafts can be made while the sap in the stock is rising.

Even when all the procedures are properly followed and all the conditions faithfully met, certain grafts take and others do not. No one really knows why this should be so; it is one of the mysteries and enticements of grafting. In general, however, the more closely two plants are related, the more likely it is that the graft will succeed. Grafting within a species almost always works, while combining different species works only some of the time. You can, for example, usually graft lemon, lime and grapefruit onto an orange tree for a limited imitation of the fanciful *table d'hote* tree. You can also graft a marianna plum scion onto a peach rootstock, but not the reverse: certain grafts between species will take one way but not the other. Between genera, the next grouping above species, grafting rarely works, and between families of genera it is virtually impossible.

When a graft takes, it usually takes with a vengeance, producing a union at least as strong and often stronger than the rest of the plant. A solidly grafted tree subjected to a high wind is more likely to break above or below the union than at the union itself. Paradoxically, however, the two parts of the plant retain their separate identities. The stock may cause the scion to grow faster or slower, larger or smaller, but the genes of the two parts will not intermingle. Pears produced on branches grafted onto quince stock may become larger but they will still look and taste like pears.

When a graft fails, on the other hand, the point of the union, as might be expected, is weakened. The failure may show up immediately—the scion dies—or it may not be evident for years. Occasionally a tree 25 to 30 feet tall, grafted many years before and showing every sign of vigor, will suddenly snap in two at the graft. Professional nurserymen blame this on the simple fact that the two plants were, after all, not meant for each other. Sometimes this problem of incompatibility can be solved by grafting a third element between

COMPATIBLE RELATIONS

the stock and the scion, one known to be compatible with both. This third element is called the interstock, and the process used to introduce it, called double-working, performs other functions as well. It may, for instance, add winter hardiness, or it could dwarf the top.

If you decide to explore the labyrinth of grafting, you will actually need very few pieces of equipment: a very sharp knife, grafting wax or a sealer such as asphalt-water emulsion dressing (if you are making spring grafts outdoors) and a half-dozen ordinary household tools. The knife is essential: it must be truly razor sharp because the success of the graft depends on two absolutely smooth, flat surfaces. Professional grafters use special thin-bladed knives that they strop constantly with leather, the way a barber sharpens a straight razor. You can use an ordinary jackknife, but you must sharpen it with an oil stone; a carborundum stone leaves it too jagged. Some woody plants, and all house plants with pulpy stems, can also be cut with a single-edged razor blade. The cut must be made in one sure, steady slicing motion; you may want to practice it on various kinds of stalks and twigs before venturing an actual graft.

A CHOICE OF TECHNIQUES   Zealous grafters are continually thinking up new ways of joining scion to stock, but most of these are refinements of a few basic methods that have proved effective through the years. Which one you use depends on the purpose of the graft, the kind of plant being joined, and the relative size of the stock and scion. Most methods fall into either of two broad categories. In one, the rootstock is split, cut or notched, the scion is inserted into the cut or notch, and the two cut edges of their cambiums meet. This type of grafting is generally done at the start of the growing season, in early spring, using scions that are still dormant. In the other category, the bark of the rootstock is pried or peeled away to expose an area of cambium, and the scion is laid against that area, the two cambiums meeting face to face. Because this operation depends on the bark "slipping"—separating easily from the wood—it is done later, in late spring or summer, when sap is flowing and growth is under way.

A particularly good method for beginners to try is the cleft graft, which belongs in the first category. Performed just as the rootstock's buds begin to swell, signaling the start of growth in the spring, it is mechanically simple and is widely used for putting new man-made varieties on fruit tree branches. It is also a good graft to use on certain house plants when the stock is considerably larger than the scion. Let us suppose you have an aging apple tree of undistinguished lineage on which you want to graft a scion or two of McIntosh apple acquired from a friend's tree or from a nursery. You should obtain the scions in midwinter, when they are dormant, taking healthy 10- to 12-inch shoots from the middle of one-year-old wood with two or more buds on each. Put the shoots in a plastic bag with a small amount of moist peat moss or wet newspaper; seal the bag and put it in the refrigerator.

In the spring, before the apple tree's buds open, at just about the time you would do the spring pruning, choose the branch on which you will make the graft. It should be about 2 to 3 inches in diameter. Saw off the end neatly, taking care not to tear the bark. With a heavy kitchen knife make a vertical split about two inches deep into the center of this stub, driving the knife down with a mallet (*page 63*). Insert a screwdriver or chisel into the center of the cut to hold it open, and withdraw the knife. Now, take your scions from the refrigerator and get ready for some nimble carving.

Note the angle of the cleft you have made, and cut the end of one of the scions so that it will fit the cleft. Try to make your cut with a single swipe of the knife, but do not worry if it does not come to an exact point. Make sure you are tapering the lower, or root, end of the scion, because a scion inserted upside down will not take. Do not touch any cut surface with your hands; oil from your fingers will prevent the parts from uniting. Set the scion into the cleft, off to one side, so that the cambium layers are in alignment. Slide it in until it

THE CLEFT GRAFT

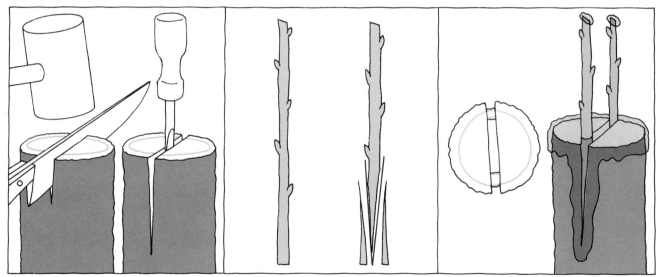

To prepare an understock for cleft grafting, cut off a stem or trunk 2 to 3 inches in diameter. With a sharp heavy knife, split the stock vertically to a depth of 2 or 3 inches, tapping the knife gently. Force a screwdriver into the cleft to hold it open.

From a species compatible with the understock, cut two scions of dormant one-year-old wood 6 to 8 inches long with 4 or 6 buds. With buds pointing upward, make two long tapering cuts (right), matching the scion to the cleft. Do not touch the cut surfaces.

Insert the scions, one at each side of the cleft, aligning the cambium layers of scions and stock. Remove the screwdriver. Cover the entire union and tops of scions with grafting wax. Rub away any buds on the stock. In late fall, cut off the less promising scion.

is snug. Then taper the other scion similarly and insert it in the other side of the cleft, again lining up the cambiums. Remove the screwdriver; the scions should now be held tightly in the cleft.

Your completed graft, with its two scions sticking up from the edges of the understock, will have to be sealed until it has safely taken. Outdoors, it is best to use an asphalt-water dressing. Or you can use grafting wax, a sticky compound that remains pliable and is specially concocted to resist freezing in cold weather and melting in hot. Grafting wax comes in either a cold, water-soluble form or in a form that must be heated. The heated variety is cumbersome to use, but it is waterproof and thus less likely to require renewing. Cover all cut surfaces with the sealer, including the cut ends at the top of the scions; make sure the cleft is completely filled, too. Within a couple of weeks you should have a verdict. If the scion's buds begin to swell, your graft has taken. Every month or so thereafter, check to make sure the wax is still intact. Late in the season or early the following year, cut off the less promising of the two scions.

On indoor plants a cleft graft does not have to be sealed. Just cover it with a plastic bag and tie the bag securely around the rootstock to keep the moisture in. After two or three weeks you can untie the bag, but if you have in the meantime moved the plant outdoors, it is best to leave the bag covering the graft loosely for another few weeks. One of the advantages of waxless grafting is that you can see the new tissue forming; by midsummer it should be well developed and you can remove the bag entirely. Keep in mind also, when grafting indoor plants, that the dormancy period of most house

THE SPLICE GRAFT

1. *For a simple splice graft, make a diagonal cut through the stock, or base of the graft. To duplicate the angle of this cut on the scion, or grafted-on section, line up the two sections side by side and nick the scion to mark the length of the cut; then slice between the nicks. Fit the stock and scion together, buds pointing upward, and wrap with grafting or surgical adhesive tape.*

2. *If the scion is considerably smaller than the stock, line it up with one side of the stock, so that the cambium layers of both sections meet. Then wrap the graft with grafting or surgical adhesive tape.*

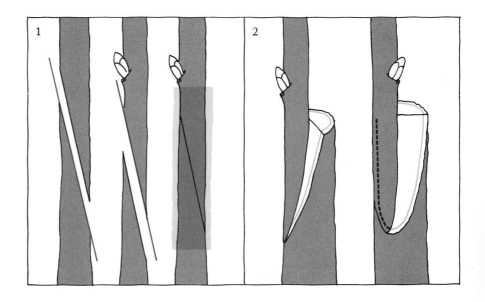

plants is unlike that of plants grown outdoors. House plants do not usually lose their leaves, even during dormancy, and in fact continue to use them to draw energy from the light. Consequently, until your scion is growing vigorously and has leaves of its own, do not remove the leaves of the understock plant, however unsightly they may be. When you feel that the scion's leaves can support the plant by themselves, you can prune off the others.

One of the other grafting techniques used early in the spring when the scion is dormant is the whip or tongue graft—whip because it is usually done on young shoots, or whips; tongue because of the way it is put together. Well suited to grafts in which scion and stock are of similar size, the technique involves making diagonal cuts of both scion and stock plus a notch in each cut (forming the tongue) to enable them to clasp together. A simpler version of this graft, using nothing more than a diagonal cut, without the tongue, is called a splice graft *(opposite)*. It works well when both plants are less than ½ inch thick; with thicker plants, however, there is a tendency for the two elements to slide out of cambial contact. To hold stock and scion together while they unite in a splice graft, wrap them with grafting tape or medical adhesive tape.

For the more difficult tongue graft, make identical sloping, diagonal cuts on both scion and stock, if possible with a single stroke of the knife, to ensure smooth, straight surfaces. A third of the way down from the tip of each, make a downward cut to form the tongue; in profile, the grafting surfaces will now look like slightly open mouths. Bring the two surfaces together, interweaving the tongues. If one element is slightly smaller than the other, place it to one side so that at least one section of their cambiums will meet. A skillfully cut tongue graft will often hold itself together, but it is a good idea to wrap it anyway with cotton string or grafting tape. If your graft is on a rootstock close to the soil, simply surround the union with moist sand or peat moss until it has healed.

A number of the grafting methods used in early spring involve joining the scion to the side of the understock rather than the top. One of them is the stub graft which, in effect, adds a new branch to a plant or tree. It is a convenient way to replace a damaged limb. The understock should be at least an inch in diameter. Using a heavy knife or chisel, make a slanting cut about an inch deep in the understock, at an angle of 20 to 30 degrees from the axis of the branch or trunk. Open the cut slightly by bending the understock back, and taper the scion to match this opening precisely. Insert the scion into the opening in order to achieve the maximum amount of cambial contact, and either tie the graft in place or nail it, using a

SIDE-GRAFTING METHODS

thin nail or two. Cover the union completely with asphalt-water dressing or grafting wax.

A useful variant of the stub graft is the side veneer graft. It works well with house plants that have stems less than an inch in diameter. Like the stub graft, it is based on a slanting cut in the understock, but the cut extends no more than a quarter of an inch into the branch or stem, and a second cut into the bottom of the first one actually removes a notched section of the understock. The first cut should be at least an inch long; the second cut should form a 45-degree angle with the first one. Cut the scion to match, with a long sloping cut on one side and a sharply angled cut on the other. Tie or nail the scion in place and protect it with a sealer.

BARK GRAFTING   When the plant material is small in size, many grafters prefer to use one of the methods from the other category, in which the bark is separated from the wood and the scion laid directly against a section of cambium. Of these, perhaps the most widely employed is the one called, appropriately, the bark graft. It bears some resemblance to the cleft graft, joining small scions to larger understock, but it must be performed in midspring or later, when the sap of the understock is running freely, enabling the bark to be easily freed.

In bark grafting the scions should be four or five inches long. They tend to perform better if kept dormant by refrigeration, though this is not strictly necessary. Start the graft by preparing the scion, tapering it to a point that slopes gently for a distance of about two inches, then ends abruptly at a very sharp angle. If your understock has thin bark, prepare it by making a two-inch lengthwise cut through the bark to the cambium. Using your knife, pry up the bark on both sides of the cut to create a space behind the bark wide enough for the scion. Gently slide the scion down into this space, with its cut surface inward, until it lies flat against the cambium.

If the understock has bark too thick to be pried up, make two parallel cuts instead of a single cut, positioning them just far enough apart to allow the scion to fit snugly between them. Remove the outer bark between the cuts and peel back the inner bark at the bottom of the cut, exposing the cambium. Now slide your scion down into this space with its pointed lower end tucked under the flap of the inner bark. Whichever bark graft you use, secure the scion to the stock by nailing it or tying it with cotton string or strips of plastic. Protect it as you would a cleft graft until the graft takes.

A second method involving the removal of bark from a portion of the plant is the approach method, in which two plants are induced to grow together while still existing on their own. This is a horticultural refinement of the process by which two plants or trees, growing

principles as conventional grafting, it differs in one intriguing respect: in temperate climates, the best time to bud is not in the springtime but in late summer, for experts have discovered that a newly inserted bud, having taken, will produce a healthier plant if the plant enters into dormancy shortly after grafting.

With budding, perhaps more than with any other form of grafting, it is important to employ the best possible understock. Rose budders often use a multiflora rootstock, which is strong and disease-resistant. For some peaches and pears the best understock is a seedling, while apples are likely to do best in rootstocks propagated vegetatively. Many years ago a celebrated series of apple rootstocks was developed by England's East Malling Research Station. These are now widely available in the United States, especially where fruit-growing is a major activity. One of the most famous is Malling 9, which has a strong dwarfing influence on a scion. These preferred rootstocks can be obtained from some nurseries, from mail-order suppliers and from some state agricultural experiment stations. Your county agent can give you advice on how to get them.

Whichever rootstock you use, it should be firmly established in the ground by midsummer, so it will be ready to nurture the new bud. Seedlings will have been started the previous year, but purchased rootstocks should have been planted in early spring, so they will benefit from a full season of growth. If the summer is dry, irrigate the rootstock several times before the budding operation begins. The success of budding depends greatly on the ability of the bark to slip; if it is dry and tight, buds do not unite well.

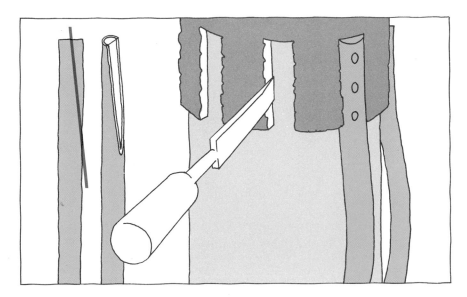

BRIDGING AN OLD TREE

Unlike the bark of a young tree (opposite), that of an old girdled tree is thick and rigid; bridging branches, therefore, must be laid into slots cut into the bark. Trim torn bark to healthy tissue. For bridges, cut pencil-thin one-year-old branches from the same tree, each 4 to 5 inches longer than the girdled area. Using the scions as patterns, cut slots into the upper and lower edges of bark at 2- to 4-inch intervals. Then remove buds and taper both ends of the scions. Insert them, top ends up, into the slots. Tack them in place and coat the graft with tree dressing.

Collect your grafting buds by cutting young shoots from the plant you want to reproduce, if possible on the very day you intend to use them. (In this instance the scion is not dormant.) Each shoot, or bud stick, as it is called, should contain a number of vegetative buds—not flower buds. Prepare each of these bud sticks by first removing all the leaves but not the leaf stems; these should remain on to serve as handles. Put the bud stick aside, preferably in the shade so it will not dry out, while you prepare the stock.

The grafting cut in the rootstock is T-shaped and gives its name to the method, T-budding. The cut should be made eight or ten inches above the soil line if you are working with fruit tree understock, two or three inches if it is a rose. First make a vertical cut a little more than an inch long through the bark to the wood. Then cross the T by making a horizontal cut at the top. Pry up the edges of the bark. Now cut one of your buds from the bud stick. First cut through the bark half an inch above the bud, then slice horizontally underneath the bud from below, taking just a bit of sapwood and using the leaf stem as a handle. Insert the bud in the understock, facing out, and push it down as far as you can, until the bark at the base of the T begins to split. Wrap the union with plastic film or with rubber bands cut into single strips, leaving the bud and its leaf-stem handle sticking out. Bud grafts need not be waxed. In a few weeks

A BUD FITTED TO A "T"

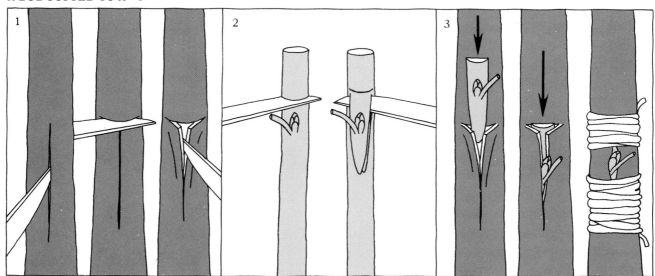

To graft a desirable bud onto hardy young stock with thin bark, prepare the stock by making a 1½-inch vertical cut through the bark. Cross it at the top with a horizontal T cut. Then pry up the edges of the bark.

From a year-old branch, choose a vegetative bud for grafting. Pinch off the adjoining leaf, but not its stem. Nick the branch horizontally above the bud and slice under it, removing an inch-long wedge with a bit of sapwood.

Using the leaf stem as a tiny handle, push the bud down into the T-shaped cut until it is completely enclosed by bark. Bind the graft with cotton string, leaving the bud exposed. When the leaf stem falls off, the graft has taken.

the handle will perform an additional service for you: if it shrivels but stays attached to the bud, your graft has not taken, but if it shrivels and falls off, you have succeeded. Plastic binding strips should be removed at this time, but if you are using rubber bands, they can stay on another month, or until they disintegrate.

Throughout the budding operation and for some months thereafter, leave the top on the understock to nurture the plant with its leaves. It should stay on until early in the following spring. Then remove the top of the understock with a slanting cut four or five inches above the bud. Use this short remnant as a stake, tying the new shoot to it until it has become well established. When it has, remove the stub just above the shoot. During this period you must also rub off latent buds that sprout above and below the new shoot.

BUDDING IN A PATCH

In areas where the growing season is longer and plants do not go completely dormant, budding can be done at other times of the year, the best being June. New growth from buds inserted in June is generally well enough established by late autumn to survive the winter. When budding is done in the summer, be sure to choose a bud stick that has ripe reddish-brown buds. The same method, T-budding, can be used, but often a slightly different technique is adopted. This technique, which takes advantage of the understock's thicker summertime bark, is called patch budding. Instead of a T-cut, you will make a square-cut opening in the bark that matches the square of the bud piece. So the first thing to do is estimate how big your bud piece will be. Cut a square this size through the bark of the understock, exposing the cambium. Then return to the bud stick, cut your bud square and fit it into the opening in the understock. In removing the bud piece from the bud stick, slide it gently sideways and press it lightly; this should separate it in such a way as to expose its own cambium layer. Apply the bud to the understock and tie it well. You may have to trim the edges of the understock's opening to line up the two surfaces.

SEPARATE IDENTITIES

Whenever grafting is done near the soil line, you will have to be on the lookout for two types of unwanted growth. If the union is close to or touching the soil, there is a chance that the scion may put down its own roots, bypassing the understock. If the purpose of the graft has been to quicken or strengthen the scion's growth, this will be of no concern. But otherwise all such roots should be trimmed off quickly. This is especially true if the purpose of the rootstock is to dwarf the size of the top growth; if the scion on such a graft is permitted to root, the dwarfing effect will soon disappear. Similarly, the understock may send up its own suckers, resulting in confusion as to which is the scion, and these too should be removed. Any grafted or budded plant whose two elements must be kept from losing their own separate identities, as in the case of a hybrid rose grafted to a wild rose rootstock, should be grafted with the union clearly and permanently above the soil line.

For many gardeners the most challenging and satisfying grafting projects are those done on small plants like peonies, roses, cacti and even the utilitarian tomato. The budding of a rose is almost identical to the budding of a fruit tree, the only difference being in the choice of bud sticks: the best are those that have borne flowers. Also, with roses you must first trim off the thorns.

THE TREE PEONIES

With the creation of the spectacular tree peony, grafting virtually becomes an art. These hardy shrubs, admired for their tremendous spring blossoms, are extremely difficult to propagate from seeds or cuttings. But they respond beautifully if grafted onto the roots of the familiar herbaceous peony. The graft takes as long as three years to achieve blossoms, but the operation itself—performed in August—is simple. And the aftercare required is modest.

The rootstock, which can be a section of root cut from any herbaceous peony, should be about four inches long and half an inch thick. The scion, cut from the tree peony to be reproduced, should be about two inches long, with two buds. A slightly modified cleft graft seems to work best. Using a razor blade, cut the bottom of the scion into a wedge-shaped instead of a tapered point with the two sides of the wedge at a 90° angle to each other. Then notch the top edge of the rootstock in a matching wedge shape and fit the scion into the notch, binding the union with surgical adhesive tape. Bury the entire assembly in a cold frame, scion upward, about two inches below the surface of a six-inch layer of potting soil over a three-inch layer of sand. When winter comes, cover the soil with salt hay. Leave the grafts there for a year and a half, making sure the soil mix stays moist, and removing the hay during the warm months. During the second spring after grafting, uncover the peony plants. If you are lucky, up to half the grafts may take and be growing briskly. Transplant them to their permanent location, and by the following year they should begin blooming.

Every year in May visitors from miles around come to see the tree peonies growing in the Long Island garden of Louis Smirnow, one of the most successful tree peony propagators. The range of colors and vividness of their colors is extraordinary.

No such splendid display results from the grafting of another sort of plant, the cactus, yet cactus grafting, practiced by an increasing number of gardeners, may be an even more impressive example of horticultural legerdemain. Cacti are grafted to improve a plant's performance, to make it grow more vigorously or to bloom sooner than it normally would. The lovely night-blooming cereus, a languid performer, turns flamboyant, for example, when grafted onto a near-relative, *Selenicereus macdonaldiae*. But cacti are also grafted for fun, to create odd combinations of form limited only by the gardener's imagination.

As with other kinds of grafting, the object in cactus grafting is to line up the cambium layers of the two plants; consequently the relative size and shape of the scion and stock are a key consideration. Also important is the growing season: winter-blooming scions should be combined, for instance, with winter-blooming stock. Beyond that, the grafting routine is more or less like that for other plant material. The best season for grafting cacti, regardless of period of bloom, is spring or summer. And for the actual operation you will need the ubiquitous razor blade, a supply of ordinary rubber bands or soft cotton string for holding the elements together, plus some toothpicks or long cactus spines as additional anchors. If the cacti are prickly, you will also need gardening gloves.

Depending on the size and shape of the cacti, four different grafting methods are possible, two of them unique to cacti. One, called the flat graft, is used when stock and scion are cylindrical and roughly the same size. You merely slice off the top of the stock and the bottom of the scion, set the two together so their cambium layers are in maximum contact, and anchor them with rubber bands or string looped over toothpicks or spines stuck into the understock. Set the graft in a warm, shady place for several weeks, then bring it into the light, but do not remove the rubber bands for a month or so. When you do, pull out the toothpicks too. You need not, however, remove spines—they will do the plant no harm.

The other special technique for grafting cacti, the stab graft, is used for joining a smaller scion to a larger stock. You simply stab the side of the stock with a knife and insert the scion—but note that the stab may be made upward and the scion inserted bottom end up. Secure the scion with a toothpick.

Cacti can also be grafted by two familiar methods, the cleft graft and the so-called side graft, which, like the splice graft, unites two diagonally cut elements. The cleft graft, in fact—made by uniting a wedge-cut scion to a wedge-cut stock—is commonly used for creating one of the most popular cactus curiosities. When the bright, trailing Christmas cactus is cleft-grafted onto the stalwart, upright jungle cactus the result is a piece of fascinating exotica: a winter-blooming cactus tree whose cascading blossoms compete for attention with the glittering decorations on a Christmas tree.

FOUR WAYS WITH CACTI

F. WIESENER, SC.

Topiary, espalier and other amusements 4

Anyone acquainted with the rudiments of pruning and grafting is a likely candidate for one of the amusing and imaginative extensions of these crafts. It is good to know how to help plants stay healthy, of course, but it is also fun to use these skills to manipulate plants beyond their conventional forms, shaping and guiding them to emphasize eccentric characteristics and sometimes to seem to defy nature altogether.

One somewhat bizarre application of pruning is topiary, the art of training and trimming plants into whimsical three-dimensional sculptures, usually geometric shapes or animal and bird figures. Such horticultural art work goes back at least to the Romans, who delighted in fashioning odd creations from cypress and boxwood. Pliny the Younger described a terrace "embellished with various figures and bounded with a box hedge, from whence you descend by an easy slope, adorned with the representation of divers animals in box, answering alternately to each other." Cypress was an ideal topiary material in Rome, for it grew freely (up to 70 feet in height, permitting stupendous designs) and it withstood shearing. But it was not hardy in northern climates.

When topiary reached its ultimate expression in England in the 16th and 17th Centuries, the favorite plant was yew. English noblemen filled their gardens with ornamental mazes and parterres amid which strutted huge yew-fashioned giraffes, roosters and rabbits. One topiary collection preserved from that era, at Levens Hall in the north of England, contains chess pieces 20 feet high, Queen Elizabeth I and two ladies-in-waiting sculpted in boxwood, and a crowned lion watching over a large bird on a pillar.

Britain's topiary craze slackened in the 18th Century in a reaction against such expensive nonsense. But the art was revived later, on a smaller scale, as the dooryard gardeners of England discovered the fun of shaping shrubbery into designs. Americans

Two 19th Century visitors get an insider's view of the wonders of pruning as they peer from a tree house made from a live maple in the Italian Piedmont. This engraving appeared in a French magazine in 1841.

generally showed little interest. But with the advent of electric hedge shears, topiary has experienced a modest United States vogue, even among the most impatient gardeners.

If you would like to try it, you must first choose an appropriate plant. The most important consideration is that it be endowed with abundant dormant buds, so that new growth will start almost anyplace you cut. In addition, the plant should have small leaves or needles to keep the sculpture from having a ragged appearance, it should be evergreen to avoid a rather woebegone winter look and it should grow quickly unless you have more than an average amount of patience.

CHOICE PLANTS FOR TOPIARY

Yew remains the best all-around choice for most topiary. It is winter-hardy in cold climates and can be pruned regularly and severely if necessary. Its one drawback is a rather slow growth rate; producing a figure 6 or 8 feet tall from a small nursery-grown specimen may take eight or 10 years. To speed things up, many gardeners start with a likely looking shrub already 2 or 3 feet tall and sculpt it gradually to their fancy.

Boxwood is easy to train in regions where it is hardy; other possibilities in the South and West are myrtle and eugenia. Privet grows fast and is easily shaped, though its hardiest species are deciduous. Hemlock and arborvitae have their fans. The Disneyland topiary animals are mostly junipers. Podocarpuses, cotoneasters and pyracanthas are other possible plants, each with a distinctive texture. For small sculptures, romantic rosemary is a good choice.

All the rules governing hedges apply. Keep cutting the plants back as they grow. Prune out any deadwood. Avoid designs that leave the lower branches shaded, since they would tend to die. Remember that all plants reach for light, so the southern side of most topiaries grows faster than the other sides. Manicure your sculpture frequently to keep it full and to avoid a scraggly look.

If you plan to sculpt an established plant, look at it carefully first to see if it suggests any special form. "I gazed at the yew before me," recalled one practitioner, "and suddenly I saw that it was beginning to look like a spiral."

For more complex designs, however, it is better to start with young plants and shape them from the outset. Feel free to grow two or more plants together for stability, their stems bracing each other. Several shrubs growing close together can be combined in a single work, perhaps with four stems providing the legs of an animal. By combining shrubs you can create any of the letters of the alphabet.

Stems and branches form the framework of your design; you guide their development with meticulous pruning. One stem might support a bird's head, for example, while a strong lateral serves as the frame for the tail. Elaborate or large designs, however, require the use of metal frames, called armatures, for stability. Some garden centers sell them. Or you can make your own from ⅛- or ¼-inch rods of soft steel sold at hardware stores, bending the rods with pipe wrenches and either wiring or soldering them together. Place the frame inside a shrub (or let the shrub grow up around it) and drive the lower ends of the rods into the ground. Tie branches and new shoots to this frame with cotton string, raffia or strips of rawhide. Keep checking these ties to make sure they do not become so tight that they girdle the branches and kill them.

Do the major shaping cautiously, a bit at a time, with hand shears, using hedge shears only to trim off surface raggedness. When you cut a main branch, prune back to a point within the outline of the projected form so the new foliage that will develop quickly will mask the cut. Most topiary plants should get their principal pruning and shearing in late spring or early summer, at the climax of their most active period of growth. Never prune evergreens toward the end of a growing season; the tender new growth thus stimulated does not have time enough to harden so it can withstand winter cold. If heavy snow comes, shake it off the top of your topiary as soon as possible to avoid broken branches. Keep your plants strong and vigorous by fertilizing and watering them regularly. Pinch back or shear lightly at frequent intervals during the summer months to maintain clean outlines.

If the four or five years needed to produce a conventional topiary seems too long a wait, there is a way to create a sculpture much more quickly for display indoors or out. Working with small-leaved, fast-growing vines planted in a pot, you can produce a showpiece in a matter of months.

Because such vines—English ivy, creeping fig or heart-leaved philodendron, perhaps—cannot stand alone, they need firm supports. You can easily shape a frame from No. 8 aluminum wire, tied at the joints with thin copper wire. Mount the frame above a flowerpot, securing it by inserting the wire ends into the soil and allowing the vine to grow up and over it. If the vine trails loosely from the frame, tie it in place with florists' string to maintain a compact shape until tendrils form.

Alternatively, you can fill a frame with sphagnum moss and plant vines in the moss so they will grow out to cover the surface. In this case, you may need to cover the frame with ½-inch chicken wire to help contain the moss. A sphagnum-moss pseudo-topiary must be kept moist; mist it daily, adding a few drops of liquid

SHORTCUT SCULPTURE

fertilizer to the water occasionally. When the frame is covered with foliage, keep it neatly groomed by snipping off unwanted tips with a pair of small scissors.

A special kind of plant shaping that is practiced in Japan works well with any of the shrubs and trees used in conventional topiary work. This technique, sometimes called cloud pruning, sometimes the pompon or poodle effect, seeks to enhance the elegance of a plant by revealing the structure of its stems and branches as the foliage is pruned into clusters. Although the completed sculpture often has the look of a windswept pine, evoking traditions of bonsai (a specialized subject not covered in this book), pompon trees are not miniatures or dwarfs but full sized. On the United States West Coast, eugenias are popular subjects for this technique.

PRUNING CLOUDS

The best cloud-pruning results will be obtained if the shrub or tree has an interesting trunk or stem, or several of them. Decide what parts you want to expose, then prune away all side growth from those areas. The remaining leaf or needle areas can either be rounded into tufts or flattened into horizontal cloudlike shapes; stems and branches will be glimpsed intermittently between them.

It is prudent to execute a cloud-pruning design on paper first, to help you avoid mistakes when you start to cut. By planning the pruning well, you may be able to achieve a zigzag look that is quite intriguing. But as in a flower arrangement, the final result should seem to be comfortably balanced, not lopsided.

While topiary shapes foliage for a deliberately staged effect and cloud pruning emphasizes the structure of stems and branches, a different approach to pruning is represented by the ancient tradi-

(continued on page 83)

A president's patterned garden

In the late 18th Century, when George Washington designed the gardens surrounding his home at Mount Vernon, Virginia, decoratively pruned fruit trees, especially trees pruned flat, or espaliered, were popular among the landed gentry. "Espaliers of fruit-trees are commonly planted to surround the quarters of a kitchen-garden," wrote Philip Miller in his Gardeners Dictionary (1763 abridgment), a book Washington avidly read.

For his Mount Vernon gardens, including the kitchen garden restored in 1936, Washington usually chose dwarf European fruit trees. Some he had espaliered against brick walls and trellises. Others, like the crab apple opposite, grown primarily for their flowers, were left freestanding, although the bottom portions of the crowns were heavily pruned to permit a "see-through" view. An astute horticulturist, Washington hinted in his diaries at a tree-grafting rivalry with one of his indentured gardeners. "I grafted six of the May white-heart Cherry growing in my walk," he wrote in 1785. "And my Gardener to show his cunning, grafted ten pears."

In Mount Vernon's flower garden, a pruned crab apple tree permits a latticework view of flowers and shrubbery beyond.

A hedge of 25-year-old espaliered apple (right) and pear trees borders Mount Vernon's kitchen garden. Although the plantation's original espalier plantings produced substantial amounts of fruit, their purpose was decorative. The bulk of the fruit crop came from orchards.

A sour cherry tree in George Washington's garden is espaliered into an informal fan shape. Because fewer branches are removed, informally pruned trees produce more fruit than rigidly shaped ones. In Washington's time willow twigs were used to train trees to their wooden trellises, but here longer-lasting leather thongs are used.

A 25-year-old dwarf pear tree, espaliered into a more formal candelabrum shape, stretches across the back of the servants'-quarters wing of the greenhouse at Mount Vernon. Age and annual pruning limit its yield.

tions of espalier. In this art, a tree or shrub is pruned and trained to become a two-dimensional plant growing flat against a wall, fence or trellis. Here the branches are meticulously guided so they grow in a pattern determined by the horticultural sculptor.

The elegant but seemingly contrived look of espaliers—the name comes from the Italian word *spalla* for shoulder, referring to the support these plants must lean on—was originally a practical response to a genuine need to grow fruit in limited spaces. Although the Romans may have used it decoratively, as they practiced topiary, the first espaliered trees were probably fruit trees grown in the cramped confines of fortified medieval towns whose warring inhabitants were unable to reach their fields. Through the centuries the Europeans, with a tradition of walled gardens, developed espalier techniques to an advanced state. Brought to the United States in colonial times, espaliering became standard in the gardens of the well-to-do, as demonstrated today in the reproductions at Colonial Williamsburg and Mount Vernon *(page 78)*.

Contemporary architecture, with its bare, blank walls, invites the use of decoratively arranged plant forms, if only to break the monotony, and many gardeners have found that contriving even elaborate forms is not difficult, once the principles of pruning have been mastered. In addition to ornamenting blank walls, espaliers make graceful fences, screens and walkway borders. If an espalier is trained to grow on a freestanding trellis, it can become a decorative focal point in a garden.

While most espaliers are formal, they need not be. Gardeners starting with stiff, mature plants often create informal arching patterns suggested by the plants themselves. The essential characteristics of an espalier are its growth in a two-dimensional plane and its need for mechanical support. Within these confines, the opportunities for inventive designs are without limit.

Fruit trees are excellent subjects for espalier, particularly those that have been dwarfed to make them more manageable. Most dwarfed fruit trees bear fruit earlier than standard trees, and their fruit benefits dramatically from the increased exposure to sunlight that espaliering provides. The fruit also remains within easy reach for tending and picking. Apples and pears are the classic fruit trees used for espalier, and to ensure fruiting they need to be planted where they will receive at least six hours of sunlight a day. But there are many other trees and shrubs that respond to the stringent espalier restrictions. Depending on your climate, you might try pyracanthas, Japanese maples, flowering quinces, flowering dogwoods, cotoneasters, junipers, yews, the smaller magnolias or for-

CHOICE PLANTS FOR ESPALIER

sythias and, in warm regions, camellias, lantanas and poinsettias. Indoors, you can attach small trellises or metal frames to pots and espalier citrus trees or such woody plants as figs or gardenias.

Outdoors, the espalier support can be any combination of wires, fencing, latticework or metal fasteners that suits the design you want. Since the traditional espalier is trained to grow closely against a wall, although not directly on it, bear in mind that a light-colored wall facing south reflects much heat in summer, especially where temperatures climb frequently into the 90s. An east- or west-facing wall may be safer to use, or you may be able to darken the south wall or roughen it to make it reflect less heat. In any case, the espalier support should be set at least 6 inches out from the wall to allow air to circulate behind the plant.

TRAINING ON WIRES   Perhaps the simplest support system consists of a series of horizontal wires 12 to 16 inches apart with the lowest 16 to 20 inches from the ground. Vertical or diagonal guiding can be achieved with thin wooden sticks or bamboo canes fastened to the wires. The plant will be trained by being tied to both wires and sticks. The wires can either be mounted on the wall or stand free on their own supports. You can use either 14-gauge galvanized wire or, better, plastic-coated marine tiller cable, available at marine-supply stores, which will chafe the branches less.

To mount the wires or cables on a wall, use metal eye bolts that can be screwed into a wooden surface or into lead sleeves inserted in holes drilled in masonry. Be sure the eye bolts are firmly set: the wires must be kept taut. A freestanding framework of wires can be strung on 4-by-4-inch posts or 2½-inch pipes; tighten the wire with turnbuckles. Make sure the wires are absolutely level.

Some nurseries sell young espaliered trees, but they are costly. With care, patience and a knowledge of pruning principles, you can easily start your own. If you already have a compactly growing tree or shrub that seems to lend itself to training in a two-dimensional plane, you may be able to adapt it to an espalier form. But shaping its branches is likely to be more difficult than with a young plant, and it may lack good foliage near its center, a hallmark of espaliers.

Starting with a one- or two-year-old whip will be easier. By controlling and directing its growth with skillful shaping and pruning, you can make it conform to any number of patterns. As with any pruning, you can speed vertical growth by removing side shoots. You can turn one stem into many by cutting back to force branching. And you can slow or halt growth by turning shoots downward or stimulate it by turning shoots upward.

Before starting, decide which pattern you want to follow (*page*

*87); changing a pattern in mid-train is difficult. Also, bear in mind that patterns differ in the kind of maintenance they will require. The more a branch is bent from its natural upright position, the more you will have to prune it. A branch trained to grow horizontally will, for instance, constantly put out new shoots from its upper surface in an attempt to restore its normal vertical growth. The simplest kind of formal espalier, and the basic form from which all others derive, is a single stem called a cordon, from the French word for cord or string. These single stems may be trained to grow vertically or diagonally and can be repeated in multiples. But assume you have a two-year-old whip to espalier and want to create an elaboration of the cordon, a form called the multiple horizontal T in which a number of branches grow horizontally from a central stem. Plant the whip and cut back its leader at a node 2 to 4 inches below the lowest wire on the framework. Several new shoots will develop at or near that cut over the next few months. Choose three, one to become the new leader and two others to form the lowest lateral branches; remove any others. As these shoots grow, prune off any side branches that develop (unless it is a tree that will bear fruit on spurs, in which case the side branches must be saved). Keep pinching back the tips of all three shoots to force them to grow slowly. As the lateral branches grow, begin tying them to the first wire, using raffia, cotton string or some other soft, flexible material. (With a spur-producing fruit tree, treat the side shoots as described on page 87.)

When the leader reaches the second wire, cut it back again 2 to 4 inches below the wire to produce three new shoots, and proceed as with the first tier. When your espalier reaches the top wire, cut the leader back to 2 inches above the third-tier wire. Then choose two shoots to form the top branches, tie them down to the wire and cut back all other growth there.

To develop a more ornate form of espalier in which the lateral branches grow out horizontally and then turn up vertically to create a candelabrum effect, fasten bamboo canes or wooden sticks to parallel wires at the points where you want vertical growth to occur. When the lateral branches reach these points, use soft ties to guide additional growth upward. Conversely, if you merely want a low hedgelike espalier along a single horizontal wire, choose two side shoots at the level of the wire to train, one in each direction, and prune away all other shoots. The shoots you have chosen will not necessarily grow out at the same rate. If maintaining branches of equal length is important to you, as is often the case when you are creating a formal design, tie the tip of the longer branch into a loop for a short while. This will slow down its rate of growth *(page 88)*.

MAKING CANDELABRA

Espalier ways and means

Espaliering requires more than just the training of plants along a two-dimensional surface; you must also make frequent and intelligent use of your pruning shears. Depending on the design and purpose of your espalier, branches growing too fast will have to be slowed down, unwanted shoots removed and others encouraged. Espaliering will be successful only when you take into account the growing characteristics of the tree or shrub (bottom left). If you use a tree that bears fruit on spurs, prune its side shoots to produce a balanced crop of fruit on all its branches (right). Working with a dwarf variety is easiest because you will not have to cut back the tree continually to control its size, yet you can prune and harvest without a ladder.

GUIDING GROWTH IN STAGES

  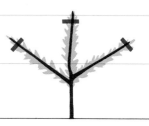

Plant a one- or two-year-old tree of a small species. To encourage multiple branching, cut about 6 or 8 inches below a taut horizontal wire 20 inches off the ground. Remove any branches below the cut.

Three or more shoots should appear just below the cut. Select one for a leader and two for laterals; remove others. Tie laterals to guides 45° from the horizontal. When tips pass the wire, pinch them back.

When the leader reaches a second wire 14 to 16 inches higher, cut it 2 to 4 inches below the wire so new branches will form. Pinch back the first laterals after every 8 to 10 inches of growth to encourage side shoots.

THE ANGLE ON DESIGNS

Some of the more common formal and informal designs for espaliering are shown at right. All are based on the modification of branches. Normally, any tree or shrub reaches toward the sun under the guidance of a hormone produced at every branch's growing tip, which is traditionally called a cordon. The more a cordon is bent from the vertical, the slower its growth will be. If espaliered horizontally or bent downward, as in the arcure style, the hormonal dominance of the growing tip ceases almost entirely. Strong shoots will sprout along the top of the branch, necessitating constant pruning. In arcure, however, one shoot is allowed to grow from the upper portion of the arch, thus channeling energy and lessening the number of offshoots.

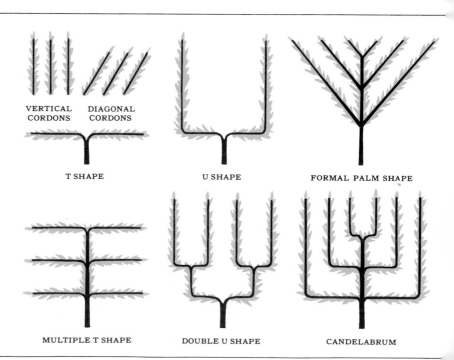

VERTICAL CORDONS   DIAGONAL CORDONS

T SHAPE

U SHAPE

FORMAL PALM SHAPE

MULTIPLE T SHAPE

DOUBLE U SHAPE

CANDELABRUM

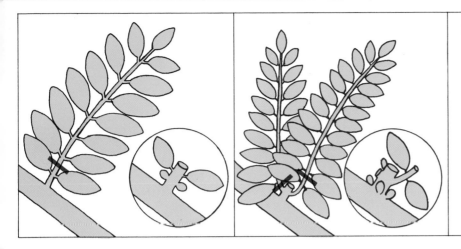

BRINGING FRUIT TO BEAR

To develop a fruiting spur on an espaliered tree, cut a side shoot back to two leaves when it is about 10 inches long (far left). Two side buds will form, remaining dormant until the following spring (inset). Then two shoots will spring from the buds. When they are 6 to 8 inches long, remove one and cut the other back to two leaves (near left). Several buds will form, and some should be flower buds that will produce fruit (inset). If not, again allow new shoots to grow out and cut them as before.

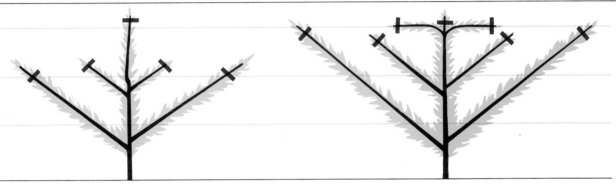

As shoots appear on the leader just below the second wire, select three and tie them as before. Continue to pinch lateral tips after 8 to 10 inches of growth. Cut the leader when it grows 2 inches past the top wire.

Select two shoots just above the top wire to be trained horizontally; cut off all others. Tie the shoots to the wire. These branches will grow slowly, but they too should be pinched back after 8 to 10 inches of growth. Pinching forces the formation of side shoots along any framework branch and will thicken and strengthen it at the same time.

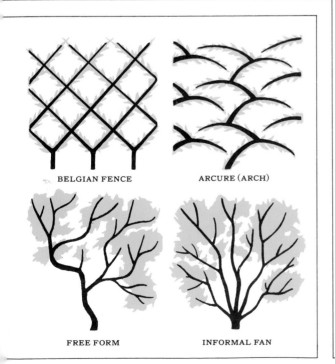

BELGIAN FENCE   **ARCURE (ARCH)**

FREE FORM   **INFORMAL FAN**

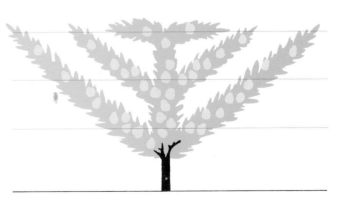

When the tree or shrub has formed the desired pattern, continue pinching the tips of all branches occasionally to slow their growth. Cut off any new growth that would mar the symmetry of the design. If the espalier is a tree that bears fruit on spurs, prune the side shoots to force the formation of spurs (top illustrations). Remove guides when the branches are strong enough to support themselves.

To create an informal fan-shaped espalier, perhaps the easiest for an amateur, keep the pairs of laterals growing in their diagonal direction, tying them in place where they cross a horizontal wire. If you like, you can cut these diagonals back in order to subdivide them into additional diagonals. The wood or bamboo supports used to create any of these patterns can be removed whenever the branches are strong enough to hold their shape without help.

ADVANCED PATTERNS   Among the more elaborate espalier designs are the Belgian fence and the arcure (arch) patterns. Both require more than one plant, and neither is suitable for dwarf trees. The Belgian fence, to be most effective, should have at least five trees planted in a row, 18 inches apart. Cut back the trees at the lowest wire and save two side shoots on each, training them along diagonal bamboo canes reaching to the top wire. As these laterals cross each other, they will form a series of decorative diamond patterns.

The arcure pattern is begun by planting a series of whips about 3 feet apart, each leaning at a 45° angle to the right. Let them grow until they are 45 inches tall, then take the end of each, pull it down, and fasten it to the base of its neighbor to the right, making a series of arcs. Because they are turned downward, these shoots will extend no farther, but each will put out new shoots along the top of the arc. Choose one strong centrally located shoot atop each arc, remove the

TECHNIQUES FOR CONTROLLING GROWTH

To slow the growth of an espaliered shoot or branch, for the sake of symmetry or to concentrate energy into the production of flower buds, bend down the top of the shoot into a loop and tie it with string.

To induce a bud to produce a new shoot, use a sharp knife to notch the branch just above the bud. The notch should remove the bark, cutting off the hormone supply that inhibits the emergence of lower branches.

To bend a branch—in this case 90°—and lead it upward between two horizontal wires, tie bamboo cane between the wires where you want the branch to be. As the branch grows, tie it with string to the bamboo.

others and let the new shoots grow to 45 inches. Then bend them to the left and tie them there. Repeat the process with a new series of shoots from the top of the arcs, bending them not to the left this time but to the right. Continue in this manner, letting the shoots grow and bending them in alternate directions, until you reach the desired height. As each arc will be about 20 inches high, supporting wires should be spaced 20 inches apart. You can frame a Belgian fence or arcure planting with a vertical cordon at each end.

There are two other decorative pruning techniques, both somewhat startling, called pollarding and pleaching. Pollarding is the annual pruning back of tree limbs virtually to the trunk, sometimes taking the entire top off, to produce a globelike display of vigorous new shoots. It is practiced on such trees as willows, lindens, poplars, beeches and the dwarf umbrella catalpa. It is more frequently seen in Europe than in the United States—the French find it a convenient way to keep street trees within bounds.

The mark of good pollarding is a rounded rather than a flattened top. One way to achieve this with a tree that has lateral branches about 8 feet up is to cut all the branches off about 2 feet from the trunk during the dormant season. In the spring, each branch will produce a series of new shoots. That fall, cut these off too. The branches will callus over, forming knoblike growths at their ends. Out of these calluses, new shoots will come the next spring. If you repeat the process for several years, the knobs will become quite large, providing an odd and unusual effect.

The other technique is pleaching, in which tree branches are interwoven and plaited together to make an arbor or canopy. To make an arbor of flowering trees, choose species that have tough but supple branches, like hawthorns and golden chain trees, planting them at about 4-foot intervals along a pathway. Let them grow until the branches are high enough to start a canopy, pruning away any extraneous growth. To guide them, build a framework of wood or pipes, supporting each tree trunk and making an arch overhead. As the branches grow into each other, interlace them until you have a full canopy of shade. Eventually the trees will support each other and you will be able to remove the framework.

Many Old World gardens employed pleaching to add the crowning touch of shade to the other elaborate horticultural delights of floral fragrance and color. One such arbor that still remains is the great tunnel of pleached ornamental lime trees that surrounds the sunken garden at England's Kensington Palace. Strolling through this fragrant, leafy corridor, a present-day visitor may easily imagine himself a member of royalty in days gone by.

POLLARDS AND PLEACHING

Sculpturing with pruning shears

Making fanciful images from living plants is a very ancient art. For centuries, the Japanese have pruned evergreens to reflect moods or lofty thoughts; sometimes they abstractly represented mountains, cascades or other elements of a beloved landscape. Gardeners of ancient Rome, working on more literal themes, carved entire fleets of ships or hunting scenes from boxwoods and cypresses.

But throughout Europe, plant sculpture was most popular during the late Renaissance when the formal gardens of royal palaces and country estates became elaborate playgrounds equipped with living toys. Menageries of animals were carved from shrubs, and hedges were laid out in artful mazes. The conceits of some of these gardens led poet Alexander Pope to report mockingly that he knew a gardener who offered for sale such sculptured items as "Noah's Ark in holly, the ribs a little damaged for want of water."

Today, such grandiose plant sculpture is seldom practiced except at historic restorations or amusement parks. But with imagination and patience, modern horticultural sculptors can revive this unusual art on a smaller scale in their private gardens.

There are several forms of plant sculpture from which to choose. Topiary, the trimming of plants into three-dimensional geometric or animal shapes such as the three bears opposite, is perhaps the best known and most demanding form. Pleaching, the braiding and pruning of closely planted trees or shrubs so their branches entwine and form a wall, tunnel or canopy, also produces an arresting effect. Ornamental hedges, planted and clipped in geometric patterns, can intrigue the eye when viewed from a higher level. And a gnarled old evergreen or crab apple tree, thinned to reveal the structure of its trunks and limbs, can be as stimulating to the mind as one contemplated in a Japanese tea garden.

Plants for sculpturing should be chosen carefully. A tree that has naturally slender branches, such as a hornbeam, would make a good choice for pleaching. And a shrub such as a yew that will not be harmed by heavy clipping is an obvious choice for topiary.

Three topiary bears add a storybook touch to a Southern California garden. Clipped from eugenia shrubs, the bears needed four years of pruning to reach this shape.

Walls of interlocked branches

The intergrown foliage atop a xylosma hedge appears to be delicately balanced on long wooden fingers. To keep its sweeping horizontal shape, the top of the fast-growing hedge of tropical evergreens is sheared every six weeks. However, the bare lower branches need only infrequent pruning to keep them free of any unwanted new growth, thus maintaining the appearance of traditional pleaching.

Rescued from a Virginia creek bank, three sheared hornbeam trees form a raised rectangular wall above a perennial flower garden. Although difficult to transplant, the slow-growing hornbeam can be kept flat on top and sides with little pruning. In Europe, its strong but supple wood is often braided and twined into more formal pleached arbors and canopies.

Illusions of clouds

Small cloudlike tufts of leaves seem to float on the tips of the branches of a sawara false cypress. This type of sculpting, in which a tree or shrub is opened up in the center to reveal its interior structure, has been practiced for centuries in Japan. False cypresses are an excellent choice for such cloud pruning, for they tolerate heavy clipping quite well.

An evergreen olive tree, pruned to expose a framework of furrowed limbs, provides decorative shade for a Southern California home. The tufts of foliage that rise from the end of each branch are sheared every six to eight weeks to keep their domed shapes. In the spring, flowers are clipped off, along with some leaves, to keep the tree from bearing fruit.

Gardens of novel delights

A playing-card garden, with each suit immaculately clipped
out of boxwood, offers a genteel reminder of the stylish
and often elaborate hedge designs once commonly found on
formal country estates. Boxwood was a popular choice for such
hedges because it could easily be pruned to hold any shape.
The four suits in the card garden above were cut from common
boxwood, a species that can live for centuries.

A statue of Pan with his magical
pipes adds an enchanting accent to
an 18-foot-high wall of Monterey
cypress trees. Native to Southern
California, Monterey cypresses grow
fast and are easily shaped into hedges
or topiary figures, for they can
withstand heavy and frequent shearing.
Left unpruned, the trees can reach a
mature height of 75 feet.

A whimsical assortment of topiary figures in a Virginia garden, mostly boxwood and yew, includes peacocks, spirals, lollipops and

a running fox. Before being brought to the garden for training, many of the boxwoods had been growing on abandoned farms.

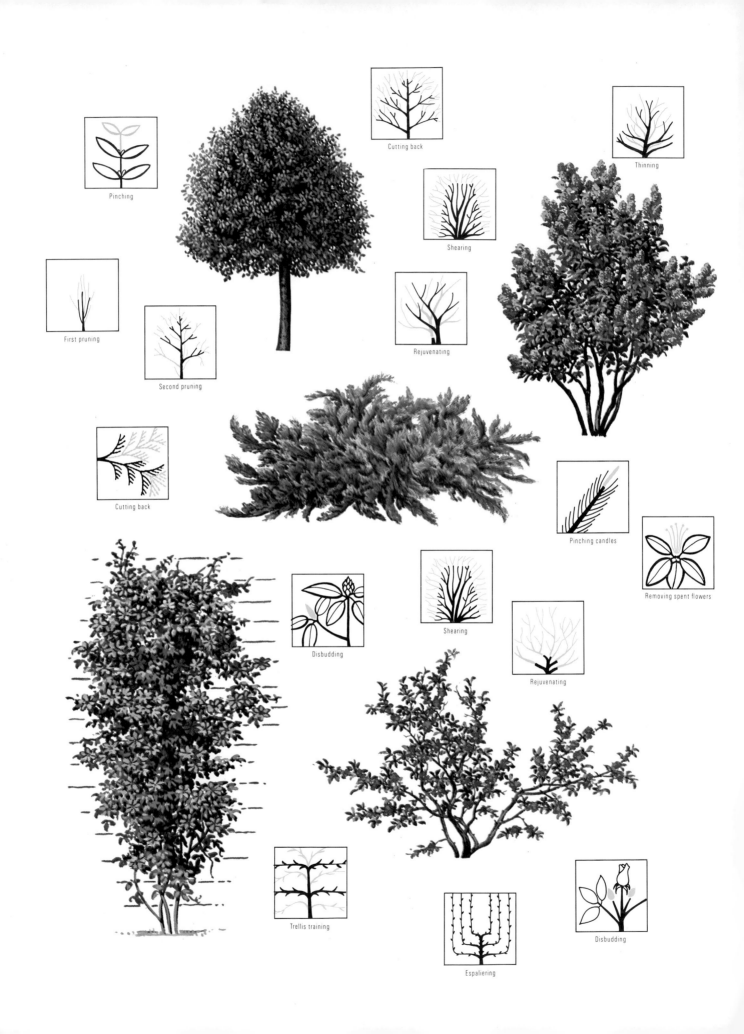

Pinching

Cutting back

Thinning

First pruning

Second pruning

Shearing

Rejuvenating

Cutting back

Pinching candles

Removing spent flowers

Disbudding

Shearing

Rejuvenating

Trellis training

Espaliering

Disbudding

An illustrated encyclopedia of pruning 5

Pruning means more than simply cutting branches off a tree or shrub to keep it from getting out of bounds. Pruning maintains the health of a plant and improves its appearance. A fruit tree is thinned to allow air and sunlight to reach the fruit in its center. The faded flowers of a lilac bush are removed to ensure blossoms the following year. The candle-like side shoots of a pine tree are pinched off in the spring so other side shoots may grow, thickening the foliage. An old, overgrown forsythia is cut back to within a few inches of the ground to force rejuvenating growth.

Plants can also be pruned to produce striking effects in the garden. Some can be espaliered so they grow flat against a wall or fence. Others can be sheared into fanciful topiary shapes. Still others can be trained to form shady canopies of foliage.

The various pruning techniques for 160 woody garden plants, including vines, conifers, broad-leaved evergreens and deciduous trees and shrubs are detailed in the encyclopedia that follows. Diagrams, like the 16 opposite, accompany each drawing. Use the diagrams not as how-to-do-it illustrations but as indications of general pruning techniques that apply to the plants illustrated. For specific pruning instructions, refer to the written entries.

The group of diagrams with each plant depicts the pruning concepts of greatest importance for that plant. For example, the diagrams for the hydrangea apply to its common shrub form, although it is sometimes trained to grow on a single stem.

Some of the groups suggest several pruning choices. A boxwood, for example, can be sheared into a formal shape, as indicated by the shearing diagram that accompanies its drawing, or it can be allowed to develop a natural, billowing, rounded shape with judicious pruning rather than shearing. Privets can be kept sheared as a compact barrier hedge or left to develop naturally with only occasional trimming of wayward branches. The choice belongs to you.

The plants and diagrams in the montage at left reflect some of the many pruning techniques described in the pages that follow. At center is a juniper, which responds well to heavy pruning.

GLOSSY ABELIA
Abelia grandiflora

WHITE FIR
Abies concolor

ABELIA

The gracefully drooping multiple stems of abelias usually need only light pruning to control their size and shape. Evergreen in mild climates, these shrubs may turn bronze in colder areas during the fall, with foliage clinging for much of the winter. Abelias are sometimes used as sprawling informal hedges and can also be sheared into formal shapes.

Abelias bloom from early summer until frost on the current year's growth. Thin after flowering to reduce the size of a shrub. To rejuvenate an old plant, cut back as much as one third of the old stems to the ground each year. In spring as new growth begins, thin out dead twigs and old growth that is slow to bud.

In colder areas where abelias are subject to winter damage, more severe pruning may be needed. In early spring as buds begin to open, thin out deadwood. Badly damaged plants can be cut to the ground; some northern gardeners routinely cut abelias to the ground each fall and protect them with mulch during the winter.

ABIES (fir)

The conical shape of this slow-growing evergreen changes slightly with age: older trees may lose their lower limbs and expose more trunk. As young trees, they need some training to maintain a dense, compact growth, but when older they require only the removal of dead or diseased wood.

Prune fir trees in late winter or very early spring when resinous sap flowing from the cuts will seal them. If the central leader, or topmost tip, is damaged, replace it by splinting and tying the nearest side branch to an upright position. Cut back branches that are too long to just above a bud or a small branch. To make the tree more compact, pinch out the center bud in the group of buds at the tip of the branches. Or wait until the bud develops into a candle, or new shoot, then snap it in half.

ACACIA

The fast-growing evergreen acacias grow as single-trunked trees or multistemmed bushes or hedges, depending on the species. These subtropical plants flower from late winter through spring outdoors in southern parts of the Gulf States and the Southwest and in northern greenhouses. Species shaped into trees, such as Bailey acacia, often suffer wind damage because their branches are brittle, while some shrub and hedge plants, such as knife acacia, may die back in cold winters. All forms are best pruned right after flowering but tolerate cutting at any time.

When young, an acacia grown as a tree must be staked and pruned for at least two or three years, until the trunk becomes strong: remove lower branches and retain one leader. Cut back new growth on the young tree to give it the desired shape. When the tree is mature, thin it every three to five years to remove crowded branches; this lessens wind damage and admits light.

On acacias grown as multistemmed shrubs or informal hedges, cut back branches of plants several times a year to encourage compact, bushy growth. Shrubs and hedges that suffer severe winter damage can be rejuvenated in spring by cutting them back to 8 to 12 inches above ground level. Shrub acacias can be sheared for a formal effect.

ACER (maple)

Shapely trees in their natural state, maples require little more than maintenance pruning when mature, but because their sap flows heavily in late winter and early spring, they should be pruned in midsummer or early fall to avoid unsightly bleeding.

Train young shade maples such as the red, sugar, Norway and silver maples to one leader, or trunk. Remove lower branches for headroom and cut back the side branches to about one third to one half their length. Once trained, most maples require only occasional pruning to remove dead or crowded branches. Soft-wooded maples such as the silver maple, however, can become thick with brittle branches that break easily in heavy winds or ice storms; they may require thinning to remove branches that have narrow crotches.

Small ornamental maple trees such as the paperbark and trident, and shrub maples such as the Japanese and Amur species are usually trained to several leaders, or trunks. Remove branches from the lowest third of the trunks of these species to exhibit the decorative bark. Thin out the branches to open the center of the tree. To maintain shrub forms, remove branches that grow vertically, and pinch the tips of the leaders to develop the strong horizontal lines typical of shrubby maples.

ACTINIDIA (kiwi berry, Chinese gooseberry)

Twining vigorously around the slats of a trellis or other support, the tender kiwi berry vine grows several feet each year to an eventual length of about 30 feet. Dormant in fall and winter, the deciduous vine bears inconspicuous flowers in spring followed in summer by hairy oval fruits that taste somewhat like gooseberries. Remove old twiggy branches and deadwood in late winter or early spring when the buds begin to swell. During the summer, cut long straggly shoots back to side buds or remove them entirely. Any heavy thinning of old wood to rejuvenate a vine should be done while the vines are dormant.

AESCULUS (horse chestnut, buckeye)

Best known in the form of the broadly spreading horse chestnut tree, *Aesculus* also includes such small, multiple-stemmed shrubs as the bottle-brush buckeye. Most species bear bold spikes of blossoms at the tips of their branches in late spring or summer, followed in the fall by glossy nuts inside rough husks. The tree has a naturally graceful silhouette; its branches curve downward and then lift in a reverse curve at the tip. The shrub forms spread from underground suckers that adapt it for use as a ground cover.

Buds swell early on these trees and shrubs, so prune in late winter or very early spring. Train a young tree to a single leader, removing the second leader that sometimes develops after wind snaps off the swollen terminal bud. As the tree gains height, remove lower branches along the trunk below 8 or 10 feet to accommodate their drooping line. Thin crowded branches and cut back new shoots that grow disproportionately fast. To train shrubs as a ground cover, peg their horizontal branches to lie close to the ground and remove vertical shoots as they appear. Renew shrubs by thinning out the oldest stems.

AILANTHUS (tree of heaven)

Growing very rapidly, even in adverse conditions, the ailanthus reaches maturity in 10 to 15 years. If trained to a single trunk, it will produce a large, open, well-rounded tree, but its luxuriant growth of suckers also makes it adaptable as a shrublike thicket. The plant's light, weak wood is susceptible to ice and storm damage that may shorten its life.

For a shade tree, train a young plant to a single trunk by pruning away basal suckers whenever they appear. Stake the tree for several years. Prune a mature tree in winter to

First pruning   Second pruning   Third pruning

COOTAMUNDRA WATTLE
Acacia baileyana

First pruning   Second pruning   Third pruning

SUGAR MAPLE
Acer saccharum

TREE OF HEAVEN
Ailanthus altissima

First pruning   Second pruning   Third pruning

WHITE ALDER
Alnus rhombifolia

First pruning   Second pruning   Third pruning

remove suckers and damaged or overcrowded branches. Do not prune any more than necessary; otherwise the tree will produce a large number of suckers. For a thicket-like growth, let suckers develop. When they are 10 feet tall, cut plants back to the ground early in the spring, before buds begin to swell, and feed generously. Vigorous new growth will sprout.

ALBIZIA (silk tree, mimosa)

The dainty, flat-topped silk tree, with its clusters of summer blossoms, provides partial shade and makes a pleasing lawn ornament. It can be trained to one or several trunks, and it grows rapidly, as much as 5 feet a year. But it is vulnerable to frost damage and disease, and with age it tends to lose its shape.

To train a young silk tree to a single trunk, prune away all competing stems; later, as the tree matures, thin out the crown every five years. Winter-damaged wood, the persistent silk tree problem, should be pruned out in late winter or early spring. If a young tree succumbs to frost, cut it to the ground. New shoots will grow; these need to be insulated for the first two or three winters with wrappings of burlap or heavy paper. When trees start to lose their vigor and shape, replace them. The silk tree does not heal readily, so cut with care and treat wounds with fungicide as necessary.

ALDER See *Alnus*

ALNUS (alder)

Alders generally grow in clumps, several trunks rising from the same roots, but the taller-growing species are often trained to a single trunk when they are young. Thereafter, they are pruned to remove crowded or weak branches and to bring out their natural forms: the Italian alder is shaped like a globe; the Manchurian and Japanese varieties are pyramidal; the white alder has branches that droop at the tips. Low-growing species like the speckled and hazel alders are treated as shrubs: one third of the old stems should be thinned out each winter, and so should any weak new growth. At the same time, cut back the remaining stems to encourage bushier growth. Very old shrub alders can be rejuvenated by cutting back the entire plants to ground level and fertilizing heavily. Most alders should be pruned in winter except for the suckers, which should be removed as they appear.

AMELANCHIER (serviceberry, shad-blow)

The open, slender-branched serviceberry grows either as a small, single-trunked tree or as a shrub that spreads on underground runners. It requires only training and maintenance pruning, and has a naturally twiggy growth that must be preserved to allow the plant to bear flowers and fruit. These appear on old wood, the flowers in early spring, the berries in early summer. Pruning, when necessary, should be done in late winter.

Train serviceberry trees to a single trunk when young by selecting a central leader and cutting back lateral branches to encourage vertical growth. Thereafter, prune annually to remove suckers and crossed or crowded branches. Rejuvenate serviceberry shrubs by thinning out the oldest, most crowded canes every fifth or sixth winter; then pinch back new shoots that appear the following spring. Curtail underground spreading by occasionally pruning the roots.

AMERICAN IVY See *Parthenocissus*
ANDROMEDA See *Pieris*
APPLE See *Malus*
APRICOT See *Prunus*

ARAUCARIA (monkey puzzle, Norfolk Island pine)

Though not a pine, the evergreen araucaria behaves like one. In mild-climate gardens, its widely spaced branches rise in symmetrical layers to form stately pyramids from 30 to 200 feet high, depending on the species. Araucaria should be pruned only to control its shape, and only on the most recent growth. If the tree is cut back to old wood, it will lose its symmetrical lines. Prune the tree in early spring before the buds start to swell, pinching back only lateral branches, never the central leader. If the central leader is broken, replace it by tying one of the branches nearest the top in an upright position. If the bottom branches hang so low that they touch the ground, cut them off.

ARBORVITAE See *Thuja*

ARBUTUS (madrona, strawberry tree)

The tree arbutus comes in two forms, the tall-growing madrona and the lower strawberry tree. Both are evergreen and are valued for their foliage and for their clusters of white flowers and red-orange berries. Though naturally multiple-stemmed, they are often trained when young to a single trunk. Thereafter, because their growing habits differ, pruning methods differ too.

The spring-flowering, summer-fruiting madrona is noted for its ornamental red-brown bark, which peels off in long vertical strips. It is also noted for its equally ornamental top growth; the branches of madrona twist in strange contorted shapes. When young, the madrona is sometimes pinched back to increase and exaggerate the amount of this distortion. To expose the attractive bark, lower branches can be removed, while thinning will expose the curious shape of the upper branches.

The fall-flowering, fall-fruiting strawberry tree—its fruit is the product of the previous year's growth—rises as straight as an arrow when it is trained to a single trunk and ends in a rounded, almost formal crown. To maintain this shape, the strawberry tree needs to be sheared repeatedly throughout the growing season.

Both forms of arbutus are sometimes thinned to make their flowers and fruit more visible. Both also tend to produce suckers and sprouts that must be removed. In addition, because of its contorted shape, the madrona tree may need to be watched especially carefully for branches that cross and rub against each other.

ARONIA (chokeberry)

This multiple-stemmed deciduous shrub is often planted as a screen or hedge. It produces flowers on new wood in late spring and berries that ripen in late summer; its foliage turns red in autumn. With very little pruning, tall species, such as red chokeberry, will produce top-to-bottom foliage. Shorter black chokeberry, often used as a border in front of taller plantings or along the edge of woods, requires some pruning because of its dense growth.

While chokeberries are young, prune to encourage branching; later, prune to encourage flowers and berries or to get more foliage. On the tall species, remove all but four or five well-spaced stems during the spring of the third year. To make a plant denser, cut back terminal shoots by half after flowering, and pinch tips of lateral shoots to induce more side branches. But if you grow chokeberries for flowers and berries as well as foliage, just thin out a few of the oldest branches in winter every three or four years. Confine low-growing chokeberry hedges to a single line by cutting out new shoots, or suckers, that grow out of bounds.

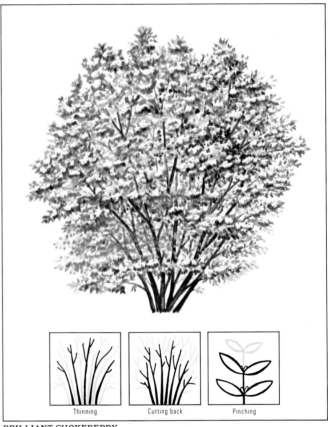

Thinning   Cutting back   Pinching

BRILLIANT CHOKEBERRY
Aronia arbutifolia brilliantissima

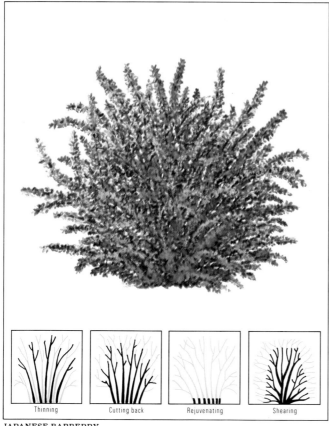

JAPANESE BARBERRY
Berberis thunbergii

ASH See *Fraxinus*

AUCUBA

These vigorous, spreading evergreen shrubs are grown in warm climates as hedge and specimen plants. If a male plant is nearby, the female plant produces decorative oval berries, usually bright red, which last all winter. Aucubas require little pruning when grown in partial shade; in full sun, the foliage tends to burn and may need to be removed. Some forms, like the Japanese aucuba, have variegated leaves with yellow markings; on these shrubs any branches bearing solid green leaves should be removed.

In early spring, prune out all winter-damaged stems. At the same time control the size and shape of the shrub by cutting back stems just above any leaf nodes to force branching at those points. To rejuvenate a badly damaged or overgrown plant, cut back the entire plant 6 to 12 inches above the ground. When they are used as hedges rather than as decorative specimens, aucubas should be trimmed a few times each summer. Use hand clippers for this job, as the plant has large, heavy leaves.

AUSTRALIAN BRUSH CHERRY See *Eugenia*
AUSTRALIAN PINE See *Casuarina*

AZALEA

Mature azaleas form graceful mounds of foliage 2 to 8 feet tall that are covered with colorful flower clusters in the spring. Deciduous azaleas generally grow upright and are sparsely branched, while the evergreen varieties tend to be denser and more spreading. If azaleas are pinched back and shaped while young, the pruning later will usually consist mainly of the removal of broken branches and deadwood (see *Rhododendron*). Prune young plants in early spring or immediately after flowering. Pinch or cut back the tips of new growth to force more branching or to keep the plants within bounds. In summer, watch for long, unbranched stems growing from the crown and cut these back to maintain a good shape. Complete the pruning cycle by midsummer for the best display of blooms the following spring, since azaleas flower on the previous year's growth.

To renew unproductive old azaleas, cut a third to one half of the oldest stems back to the ground in spring and pinch new shoots as they appear. Azaleas produce new vegetative buds along their stems wherever they are cut, so they can be trimmed wherever necessary to force new branching. Kurume azaleas can even be gently sheared with hedge shears. When shearing an azalea, follow its natural contours to avoid creating a misshapen plant.

B

BALD CYPRESS See *Taxodium*
BARBERRY See *Berberis*

BAUHINIA (Hong Kong orchid tree, red bauhinia)

Left to grow naturally, these tropical plants become multi-stemmed shrubs or single-trunk trees that bear orchid-like flowers. They are pruned when young to shape them, after the flowers fade. Because they are sensitive to frost, they may need to be pruned in the spring to remove damaged tips and encourage new growth.

Tree species such as the evergreen Hong Kong orchid tree should be trained to a single leader, or trunk, with a few lateral branches. Staking may be necessary until the trunk thickens. These species grow from 6-foot saplings to 20- to 40-foot specimen trees, and once the loose canopies of foliage are established, they need only maintenance pruning to remove dead or damaged tips.

Shrubby species with twining tendrils, like the red bauhinia, can be trained to grow as a vine against a wall or fence by annual thinning and cutting back.

BAYBERRY See *Myrica*
BEAUTY BUSH See *Kolkwitzia*
BEAUTY-BERRY See *Callicarpa*
BEECH See *Fagus*
BEEFWOOD See *Casuarina*

BERBERIS (barberry)

The multiple-stemmed barberries come in two forms, deciduous and evergreen, and are best known for their red, yellow or black berries in autumn and winter. They produce small flowers in the spring. They are planted as hedges or as specimen shrubs. Most evergreen species grow upright in naturally graceful arching shapes, but left unpruned, canes may cluster so thickly that there is no room for branches. Deciduous species are denser, making them ideal for barriers; several varieties have decorative red foliage.

Unless they are intended as hedges, allow barberry bushes to follow their normal growth patterns for the first few years. If set as a hedge, shear them to an A shape when they are small, to expose lower branches to the sun; then, when the hedge is the desired height, trim the top flat but retain the sloping sides. Thin deciduous barberries grown as specimens every second or third winter, removing a few of the oldest branches to admit light to the interior. To expose more berries for winter display, prune out some of the twiggy top growth. Rejuvenate old, unmanageable bushes by cutting them back to the ground in early spring. When the stems of evergreen species cluster too thickly to produce branches, thin out weak and old canes to the ground in winter or early spring before the buds open. Or pinch off emerging shoots at ground level in the summer before they get a chance to crowd the mature canes.

BETULA (birch)

Birches grow rapidly, reaching a height of 20 feet or more in 10 years. Their natural shape is either pyramidal or conical, depending on the species, and they may have one or several trunks. Species that have drooping or weeping branches, such as the European white birch, are generally trained to one trunk; tall species, such as the snowy-barked canoe birch, have one or more trunks; shorter species, like the red birch, grow in clumps of as many as six trunks that rise from a single root. The bark of white birches and other smooth-barked varieties is easily scarred. Although birch trees bleed at nearly any season, their sap flow is heaviest in late winter and early spring.

Birches should not be pruned any more than necessary because pruning subjects them to decay, and they should never be pruned during their heaviest sap flow. Young trees, however, can be pruned for shape. To train a sapling to a single trunk, remove the lower lateral branches before their diameter exceeds ½ inch, and remove any upper branches that might interfere with the central leader. During this training period, provide support for the trunk of a weeping birch. Remove all lateral branches that are less than 12 feet from the ground. To produce multiple trunks, cut the sapling back to the ground when it is 3 feet tall and train the strongest of the new shoots as trunks.

BIG-LEAVED HYDRANGEA See *Hydrangea*

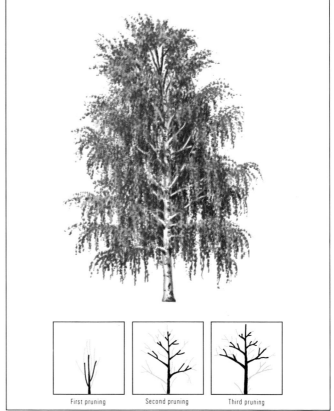

First pruning   Second pruning   Third pruning

CUT-LEAVED EUROPEAN WHITE BIRCH
Betula pendula gracilis

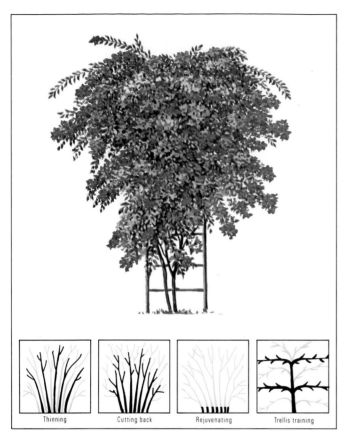

BARBARA KARST BOUGAINVILLEA
Bougainvillea 'Barbara Karst'

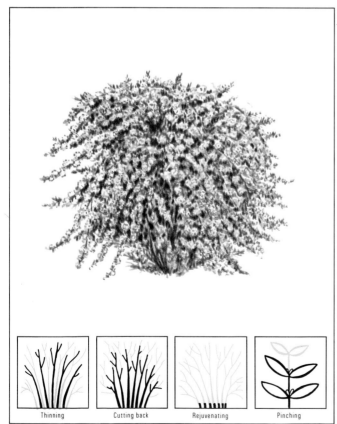

FOUNTAIN BUTTERFLY BUSH
Buddleia alternifolia

BIGNONIA (trumpet vine, cross vine)

The evergreen trumpet vine, which is thick with stalked leaflets, climbs rapidly by means of twining tendrils to blanket a high trellis or wall with year-round greenery. Clusters of red-to-orange funnel-shaped flowers bloom in the fall in some areas, in late spring or summer elsewhere. Flowering dictates the time of major pruning, since blooms appear on new growth. Remove all weak or damaged wood and thin crowded stems inside the vine for good air circulation and light. Take care not to leave stubs, since this plant is susceptible to rot. Train to a single layer of foliage by pinching out upright shoots and cutting back overgrown runners (*see Campsis*). Remove suckers from the base of the plant.

BIRCH See *Betula*
BLACK RASPBERRY See *Rubus*
BLACKBERRY See *Rubus*
BLUEBERRY See *Vaccinium*
BOSTON IVY See *Parthenocissus*
BOTTLE BRUSH See *Callistemon*

BOUGAINVILLEA

A woody tropical vine, bougainvillea is usually grown as a wall plant on a trellis; its flowers and vividly colored bracts appear on tall straight shoots that form in the spring and blossom in summer. In sufficiently warm climates, bougainvillea may grow 20 feet or more in a single year and must be kept under control by heavy pruning, sometimes twice a year. Thin out some stems before new growth starts, to establish a balanced scaffold of woody trunks evenly spaced over the trellis, and cut back year-old lateral branches to two buds each. If lateral growth is very thick, remove some branches completely or cut them back severely, leaving only short spurs where they join the main stem. Remove any suckers that appear. After the flowers have faded, remove the twiggy shoots that bore them.

In cooler areas, bougainvillea grows less rampantly, and severe pruning may encourage stem growth at the expense of flowers. Branches that may appear to be killed by frost should not be pruned out until the vine's new growth begins; often these branches revive.

Old bougainvillea vines can be rejuvenated by cutting them to the ground or as far back as seems advisable.

BOXWOOD See *Buxus*
BRIDAL WREATH See *Spiraea*
BUCKEYE See *Aesculus*

BUDDLEIA (butterfly bush, summer lilac)

All species of buddleia grow rapidly, producing as much as 8 feet of new growth in a single summer. They are multiple stemmed, and their slender, arching branches bear flower spikes 12 or more inches long. Some species, such as summer lilac, bloom on new growth; others, like the fountain butterfly bush, bloom on the growth of the previous year. Many species are badly injured or even die back to the ground in cold winters. Butterfly bushes are generally grown as specimens and are allowed to reach their normal height and follow their natural spread.

Buddleias that bloom on new growth should be cut to the ground in early spring to encourage new growth from the base of the plant. Buddleias that bloom on old growth should be pruned after they have flowered. Remove shoots containing spent blossoms and thin out any stems that are weak or spindly. Rejuvenate old plants by cutting them to the ground in the spring; thin the subsequent new shoots to reduce crowding and pinch back the tips of remaining shoots to encourage branching. For bushier plants, pinch off the tips of new growth on all species repeatedly through the summer.

BUSH CLOVER See *Lespedeza*
BUTTERFLY BUSH See *Buddleia*

BUXUS (boxwood)

Few plants take better to precise geometrical shaping than the densely branched boxwoods, but these slow-growing evergreens can also be left unpruned to form billowy masses of pungent foliage. Smaller species, such as littleleaf boxwood, may be sheared into globular or pyramidal specimen shrubs, or they may be trained as formal hedges. Larger species, such as common boxwood, can also be grown as hedges or freestanding specimens and are often sheared into tree forms or elaborate topiaries.

To shape a young boxwood into a tree, select a strong central leader, thin competitive upright shoots back to the ground and cut back side shoots. In subsequent years remove lower branches until the trunk has attained the desired height. For hedges and topiaries, pinch back newly planted stock to encourage branching, and thereafter shear closely for about three years to stimulate dense growth. When the desired shape has been established, cut back the outside branches annually in late spring before new growth begins, to keep the plant's shape compact; trim off any ragged new growth in summer. Because of its twiggy growth habit, boxwood generally requires maintenance pruning, also in the spring, to remove deadwood. In colder climates, severely damaged plants can be pruned back close to the ground to rejuvenate them, and the resulting shoots can then be pruned and shaped as new plants.

C

CALAMONDIN ORANGE See *Citrus*
CALIFORNIA LILAC See *Ceanothus*

CALLIANDRA (powder puff, flame bush)

Flowering broad-leaved evergreens, the semitropical calliandras grow from multiple stems and tend toward a rounded shape, whether they are 6-foot-tall shrubs or 30-foot trees. They may be planted to form a screen or as lawn specimens in the frost-free parts of Southern California, New Mexico and Florida. Powder-puff calliandra blooms from autumn through spring; flame bush in late spring and summer. Both species bloom on new growth and should be cut back immediately after flowering to foster new blossoms the following season.

Other than the removal of dead or crossing branches, the powder-puff calliandra requires little pruning. Flame bush benefits from thinning to preserve and expose the graceful lines of its branches. To rejuvenate both calliandras, prune some of their oldest stems back to the ground annually.

CALLICARPA (beauty-berry)

A deciduous shrub, the callicarpa grows 4 to 6 feet tall, with upright multiple stems that may die back to the ground in cold climates. Its inconspicuous flowers, which are followed by clusters of bright autumn berries, are produced on new growth. The berries last only two or three weeks after the leaves drop in autumn.

If the stems of a callicarpa survive the winter, thin out old wood and ungainly branches to encourage new growth and preserve the shape of the bush. In colder climates, cut to the ground in early spring all stems that have suffered winterkill.

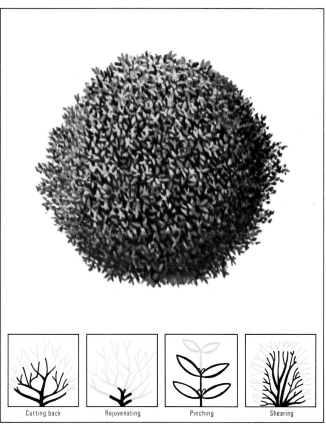

EDGING BOX
Buxus sempervirens suffruticosa

Cutting back   Rejuvenating   Pinching   Shearing

PURPLE BEAUTY-BERRY
Callicarpa dichotoma

Thinning   Cutting back   Rejuvenating

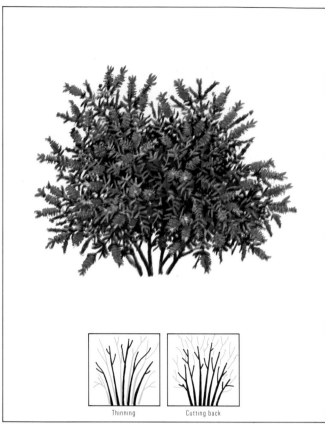

LEMON BOTTLE BRUSH
Callistemon citrinus

COMMON CAMELLIA
Camellia japonica

Callicarpa is easily transplanted, but plants that are over 4 feet tall should be cut back to half their height after they have been moved.

CALLISTEMON (bottle brush)

Blooming throughout the year in warm climates, the evergreen bottle brush grows from multiple trunks, sometimes at the rate of 2 or 3 feet a year while young. It can be allowed to grow as a bush or trained as a small tree, and can be used as a single specimen or as an informal hedge. In whatever form it is grown, bottle brush should be pruned at least once a year, after its heaviest flowering, which generally occurs in spring or summer.

To shape young plants and promote bushiness, pinch back tips and cut back the most vigorous new shoots to the part where growth is thickest. To train bottle brush as a tree, prune to a single stem and remove its lower branches. Thin unwanted branches and stems to prevent top-heavy growth. When it is trained as a tree, bottle brush must be staked for support during its earlier years.

CALLUNA (heather)

A low-growing evergreen shrub used as ground cover, heather is a densely branching plant that flowers in late summer and early fall on new growth. Remove deadwood and old flowers as they appear. To keep plants compact and full and to increase bloom, prune heather every year in early spring before new growth begins, cutting back most of the previous season's growth. Use the same technique to rejuvenate old heather plants. Dwarf varieties do not require any pruning except the removal of deadwood. All heather may be sheared if desired.

CALYCANTHUS (sweet shrub, Carolina allspice)

This vigorous, aromatic flowering shrub grows as wide as it does high and must sometimes be pruned to keep it within bounds, as well as to keep weak, dead and crowded branches thinned out. It tends to spread horizontally at ground level and to send up suckers from underground stems. Prune in early spring before new growth starts, so that the solitary blossoms, which appear in late spring at the tips of new growth, will have a chance to develop. After flowering, cut back the young shoots to shape the plant; in the fall, remove some of the arching branches.

CAMELLIA

Although all camellias are evergreen shrubs, some of them grow low and spreading; others are tall and upright. Specimen plants are generally allowed to follow their natural shape, but camellias grown as informal hedges are cut back with hand clippers to stay dense and compact.

Established plants can be pruned lightly after their blooming period in late autumn or spring depending on the variety. Remove weak branches or those that have suffered winter damage; shape plants as necessary by cutting back overlong branches to two or three dormant buds and thinning out to relieve overcrowding.

CAMPSIS (trumpet creeper, trumpet flower)

Climbing by means of aerial rootlets, the deciduous trumpet creeper vine grows so rampantly that it can quickly cover and weigh down a trellis if not pruned each year. Cut side branches back to two or three buds in late winter or early spring. Heavy pruning during the growing season will reduce the number of orange trumpet-shaped blooms that appear among the oval leaves on new growth in summer or early fall. Control the shape of the vine during the growing season by pinching and cutting trailing shoots back to two or three leaflets. Remove suckers that grow at the base of the plant as soon as they appear.

To rejuvenate an old trumpet creeper, cut it back to the ground in winter. When new shoots appear in spring, select one or two to train as the new plant.

CARISSA (Natal plum, hedge thorn)

These thorny evergreen shrubs and small trees can be trained into upright or densely mounding lines for use as foundation plantings and hedges. In warm climates, the flowers and fruit are produced year round, and pruning can occur at any time. In areas where frost may occur, plants should not be pruned after late summer.

To train Natal plum as an upright shrub, remove branches close to the ground and thin upper branches regularly to establish a well-spaced branching pattern. Cut back new growth as needed to maintain shape. For bushier growth, pinch back new shoots when they are about 1 foot long.

To train Natal plum as a formal hedge, cut back after planting to about 6 to 8 inches in height, then continue to cut back about 6 inches for every foot of growth until the hedge reaches the desired height.

On all plants, remove deadwood and thin out about one third of the old wood to prevent crowding. Badly damaged Natal plum can sometimes be rejuvenated by cutting back the entire plant to about one foot from the ground and then removing all but about five of the new shoots.

CAROLINA ALLSPICE See *Calycanthus*

CARPINUS (hornbeam, ironwood)

Hard, durable wood has earned these trees forbidding common names, but hornbeams are easily pruned when necessary. With dense foliage, they naturally assume compact forms and shape themselves with little pruning. The conical European hornbeam grows an average of a foot a year to an ultimate height of 60 feet. The American hornbeam has a broader spread and develops several trunks. Prune these deciduous trees during the winter. Train young trees to a single leader by cutting branches back about one quarter of their length at the time of transplanting. Young hornbeams can also be sheared into tall hedges. Mature hornbeams need only be pruned to remove dead or diseased wood.

CARYA (pecan)

Open-crowned pecan trees grow slowly into gigantic specimens more than 100 feet tall with thick trunks, sturdy limbs and rounded crowns of broad, deciduous leaves. Southern species of pecan trees are grown primarily for their abundant crops of smooth, thin-shelled nuts, while northern species make fine shade trees.

Pecans should be planted from late fall to early spring, and when grown for shade, they should be pruned immediately. Cut back all side branches by half their length. When the tree is 4 to 5 feet tall, select a leader, or trunk, and remove competing branches. As the tree grows, allow branches to develop about 18 inches apart along the trunk.

For nut crops, plant at least two cultivars close together to ensure pollination. Since nuts fall to the ground at harvest time, young nut trees can be trained in the same way as shade specimens. But some experts recommend a modification for best production. To modify a leader, let the tree develop normally until branches are 4 to 5 feet long. Then cut off the leader about 4 feet above the ground and remove

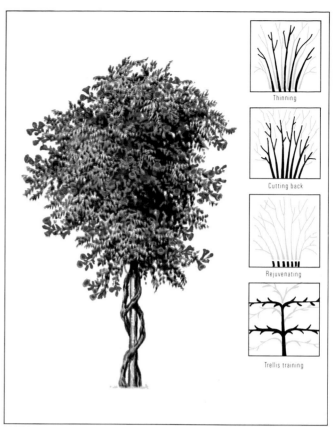

MADAME GALEN TRUMPET VINE
Campsis tagliabuana 'Madame Galen'

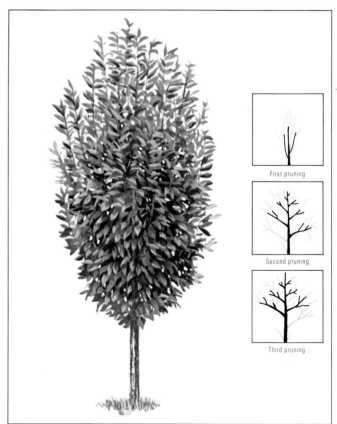

PYRAMIDAL EUROPEAN HORNBEAM
Carpinus betulus fastigiata

all but three to five strong, well-spaced wide-angled side branches. Cut these branches back several times as the tree grows, keeping them several inches shorter than the leader.

Pecans begin to bear four to seven years after planting, with nuts ripening in early fall on the previous year's growth. Prune mature trees in late winter before new growth begins, cutting back any weak or damaged limb to an outward-facing bud. The branches and twigs of pecan trees are very brittle and are therefore subject to wind damage.

CASTANEA (Chinese chestnut)

Immune to the blight that has practically exterminated the native American chestnut, the 30- to 60-foot Chinese chest-nut is grown for the shade cast by its glossy green deciduous foliage, for its small clusters of white flowers that appear in early summer, and for its burrs that break open in fall to release several tasty nuts.

With a naturally twiggy, sometimes shrubby growth habit, young Chinese chestnuts may be trained to a tree shape in either of two ways. If the young plant has a central leader, or main stem, remove scaffold branches growing from the main stem for about a foot below its tip, and severely cut back all side branches growing from the remaining scaffold branches to one half their original length. If the plant lacks a leader, choose an upright-growing branch at the top of the plant and remove nearby competing branches. Cut back side branches of the new leader by one half their length. For the next two or three years, remove any competing leaders until the trunk has become established.

The best time to prune is in the winter. As young trees grow, thin crowded branches and encourage development of wide-crotched scaffold branches by removing branches that grow from the trunk at narrow angles. Mature trees require only maintenance pruning to remove suckers, water sprouts and weak or overcrowded branches.

CASUARINA (beefwood, Australian pine, she-oak)

Fast-growing beefwoods, grown in warm climates, make handsome specimen trees or they can be shaped into screens, formal hedges or topiary figures. Raised primarily for their needle-like foliage, they can be pruned at any time, without regard for the insignificant flowers and cone-shaped fruit.

Beefwoods grown as specimen trees are trained to a single leader when young; thereafter they need a minimum amount of selective thinning to remove deadwood and maintain an open, airy center.

To shape beefwoods into hedges, cut them back regularly to encourage bushier growth. Cut back to nodes just above the crotches where side branches begin; new buds will sprout just below the cuts.

CATALPA (Indian bean, umbrella tree)

Allowed to grow naturally, the sturdy catalpas will grow from 30 to 100 feet tall, depending on the species. Although they can grow as multistemmed specimens, they are usually trained as shade trees with single trunks. The towering west-ern catalpa is pyramid shaped while the smaller common or southern catalpa is round headed. There is a grafted variety of the common catalpa nicknamed the umbrella tree because of its shape: it produces long weeping shoots that may touch the ground. With age, the branches of catalpa plants become gnarled and pendulous.

Prune catalpas in winter, when they are leafless and dor-mant, or in early spring. Pyramidal clusters of blossoms appear in early summer. To train catalpa to a shrub form, choose several leaders and cut back the branches to encour-

First pruning   Second pruning   Third pruning

CHINESE CHESTNUT
Castanea mollissima

age bushy growth. Thin the top periodically to admit light.

Train a young tree to a single trunk by choosing a single leader and removing branches 8 to 10 feet above the ground. As the tree develops, cut back branches to encourage denser growth. Cut back heavy pendulous branches of mature trees. Remove shoots of the umbrella tree annually to the same points on the old growth, thus building up thick knobs that sprout increased numbers of new shoots the following year.

CEANOTHUS (California lilac, New Jersey tea)

There are two forms of ceanothus, evergreen and deciduous. Evergreen ceanothuses are tender plants that flower on old growth. They are used as ground covers or as multi-stemmed shrubs or small trees; the shrub version is often trained against a wall. The hardier deciduous ceanothuses are bushy plants that flower on new growth of the current season. Spring-flowering species are pruned after they flower, summer-blooming species in early spring.

To control evergreen ground covers, thin out the trailing branches and cut back upright stems. To keep freestanding shrubs shapely, remove a few of the oldest stems each year and cut back new growth. Shrubs grown against a wall need more severe pruning: cut back the previous year's growth to two or three buds and cut back branches growing away from the wall to two or three leaves.

Most deciduous ceanothuses are many-stemmed shrubs, like the New Jersey tea, that require little care. Thin out one or two of the oldest stems every two or three years. To keep the shrubs at the desired height, cut back new growth to two pairs of buds in early spring before new growth begins.

CEDAR See *Cedrus, Juniperus* and *Thuja*

CEDRUS (cedar)

Conical young cedars become towering, flat-topped trees as they mature. Like other long-needled evergreens, they should be pruned lightly and allowed to take their natural shape. Prune them in late summer after sap flow has slowed and new growth is complete, taking care to cut new growth just above a bud or at its base. Larger branches cut back to old wood may not initiate new shoots.

When young, cedars often appear straggly and lack distinct leaders. Encourage a leader to develop by allowing several stems at or near the top to grow naturally for two to three years, then remove all but the most vigorous upright shoot. Once a tree is established, you can cut back new growth on side branches to shape the tree. By removing half the new growth from the tip of each branch, you can encourage thicker branching, but this pruning is not a necessity.

Prune older trees to remove lower branches for head clearance or when branches become scraggly due to shade from the growth above them. Tangled or crowded branches should be thinned by cutting the branches off at the trunk.

CELTIS (hackberry)

A rounded head of dense deciduous foliage makes the hackberry a valuable shade tree in warm climates. During each spring of the first few years after planting, train the rapidly growing young tree to a strong central leader by cutting back the side branches along the main stem, leaving each branch with only two or three leaves. Mature hackberry trees need only routine thinning to remove dead, diseased or crowded branches. Native hackberries are often afflicted by a fungus that causes unsightly tufts of twiggy growth, called witches'-brooms, along its branches. Cut off any of these growths that you are able to reach.

CALIFORNIA LILAC
Ceanothus thyrsiflorus

JAPANESE QUINCE
Chaenomeles japonica

CERCIDIPHYLLUM (katsura tree)

This deciduous shade tree generally grows with several trunks and wide-spreading branches, but it can be trained to a single trunk. It is used as both a street tree and a lawn specimen and is noted for its red and yellow autumn leaves as well as for its rapid growth, becoming 20 to 25 feet tall in half-a-dozen years.

Prune the young katsura to a single trunk if it is to be used as a street tree; otherwise let it grow with several trunks. In the latter form, the tree may require thinning to avoid crossed branches. The little pruning required by a mature tree should be done in the autumn after the leaves have dropped; thin crowded branches and remove those that appear weak. Remove dead twigs.

CERCIS (redbud)

Growing wild over much of the United States, the redbud quickly becomes a shrub or small tree with a single or multiple trunk. Clusters of purplish-pink flowers appear in spring prior to the heart-shaped deciduous leaves. Shape a young plant in late winter, before new growth begins, by thinning out competing branches and cutting back upper branches to develop a graceful spreading canopy.

Redbuds are short-lived plants and are susceptible to decay. Prune mature redbuds as little as possible, thinning them only to remove deadwood. (The Chinese redbud is generally a multiple-stemmed shrub that requires only the removal of dead or injured branches.)

CHAENOMELES (flowering quince)

Hardy deciduous shrubs, quince bushes produce flowers and fruit on the same short spurs for several years running. Taller, upright forms such as Chinese quince make good specimen plants and can be espaliered. Low-spreading Japanese quince can be used for foundation plantings or informal hedges. Once flowering begins at two to three years, the quince needs little pruning, but it should be given a well-balanced open shape at the outset. Choose several strong canes and thin out the others. To help produce spurs, cut back new growth to three or four buds in the winter. Every three or four years, thin out the weak or crowded canes of old shrubs. Remove suckers as they form around the base.

To espalier quince, fan out the canes and branches against a wall and remove all outward growth. Shorten new growth in summer to four leaves, and in winter cut back the same growth to two buds. When growing quince as a hedge, cut the plants back during the summer after they have flowered to encourage bushiness.

CHAMAECYPARIS (false cypress)

These evergreen trees, shrubs and miniature rock-garden plants are pruned according to their garden use. Most of the tall ornamental lawn species, such as Lawson cypress, need little pruning, but they should be trained to a single leader or trunk to prevent dual trunks from splitting apart as the trees age. To thicken their shape, cut back part of the new growth as it matures. Shorten long branches, but be careful not to cut back to old wood from previous years or the cut branches will die. To stimulate new growth and thus conceal unsightly dead foliage on inner branches, cut back yearly.

Shrub types are generally pruned as hedges. They should be sheared in early spring before growth starts and thereafter as needed to maintain compact growth. Slope the sides inward at the top.

Prune miniature rock-garden species of false cypress by pinching them back frequently.

CHASTE TREE See *Vitex*
CHERRY See *Prunus*
CHINABERRY See *Melia*
CHINESE CHESTNUT See *Castanea*
CHINESE FLAME TREE See *Koelreuteria*
CHINESE GOOSEBERRY See *Actinidia*

CHIONANTHUS (fringe tree, old-man's-beard)

Although the white-blossoming fringe tree may be grown with a single trunk, it is safer to let several trunks develop; the plant is susceptible to borer damage, and several trunks give it a better chance to survive. The fringe tree has a slow rate of growth and can be trained as an ornamental tree or shrub. Choose leaders to determine the shape and spread of the mature plant. Thin to reveal the graceful pattern of its branching and pinch back for bushier growth. Renew the old bushes by thinning out the oldest stems. Prune male plants after flowering in summer, but postpone the pruning of female trees until early winter after the birds have eaten the dark blue fruit. Prune borer-infested trees radically; trees with infested trunks should be cut to the ground.

CHOKEBERRY See *Aronia*

CITRUS (calamondin orange, grapefruit, lemon, lime, orange)

Though generally grown for their fruit, both dwarf and standard-sized citrus trees have glossy evergreen leaves and fragrant blossoms, making them admirable ornamentals as well. Regardless of purpose, all young citrus trees must be pruned and shaped to prevent them from becoming lanky, unattractive plants. The fruit of most species ripens in winter to spring, though some bear fruit all year round. Generally they should be pruned in spring and summer.

To train newly planted trees, remove all but three to five branches, spaced 6 to 12 inches apart, facing in different directions and forming wide angles with the trunk. Cut back the selected branches so that all are about the same length. Allow only a few more branches to form over the next few years. Cut back long shoots and thin out crowded and off-balance branches whenever necessary.

To rejuvenate old trees, cut back branches of standard trees 1 to 3 feet; cut back branches of dwarf trees to 12-inch stubs. Remove deadwood. When new growth starts, thin out about half of the new shoots as soon as they appear and another half later in the season. Pinch the tips of the remaining shoots to promote branching.

Citrus plants grown in pots indoors bear more fruit if they are kept thinned and under 3 feet tall. Pinch back new shoots in early spring and again later in the growing season.

CLEMATIS (virgin's-bower)

Prized for its open flowers in clear colors, the clematis vine has twining leafstalks that enable it to be trained easily to wires or trellises, up trees or on mesh supports. Where there is room, clematis will sprawl and cover the ground. All of these perennial plants are deciduous, but species differ in their period and pattern of bloom.

Newly planted vines should be pruned back to two or three pairs of buds. To keep plants low, cut back their vertical growth to just above the first joint of the previous year's vertical growth. To get taller plants, allow vertical growth to develop and cut back side branches, leaving just a single joint of the previous season's growth on them. Thin out excess shoots to prevent crowding.

After the vine is established, the amount and time of

First pruning   Second pruning   Third pruning

SWEET ORANGE
Citrus sinensis

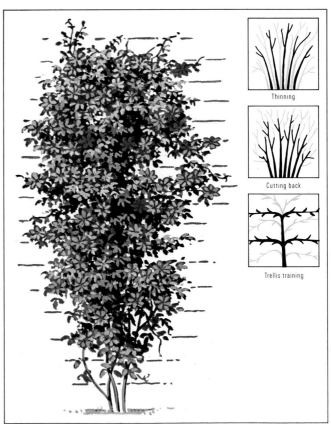

Thinning

Cutting back

Trellis training

RAMONA CLEMATIS
Clematis 'Ramona'

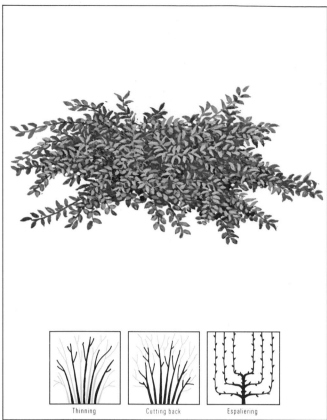

BEARBERRY COTONEASTER
Cotoneaster dammeri

WASHINGTON HAWTHORN
Crataegus phaenopyrum

pruning depends on the time of year that a particular clematis blooms. Spring-flowering clematises, which bloom on the previous year's growth, should be thinned and cut back lightly after they flower, allowing time for new growth to develop for the next season's blooms. Summer-flowering clematises, which bloom on new growth, should be cut back in fall or in early spring as buds begin to swell, removing all but 12 inches of the previous year's growth. A third kind of clematis, which flowers in both spring and summer, can be pruned lightly after it flowers in spring to get the most from the spring flowering, or pruned hard in the spring as buds swell for maximum summer bloom.

COMMON LANTANA See *Lantana*
COMMON PEARLBUSH See *Exochorda*
CORNELIAN CHERRY See *Cornus*

CORNUS (cornelian cherry, dogwood)

Depending on their form, dogwoods respond in different ways to pruning. A tree form, the flowering dogwood, heals so slowly when cut and is so prone to decay that it is only pruned for maintenance reasons. Its distinctive beauty of line, with single or multiple trunks and layered horizontal branches, is in any case one of its main attractions, so there is little reason to modify nature. This form should be pruned after flowering, since flower buds formed in the summer bloom the following spring. Remove dead or damaged wood, crowded or crossing branches, and inner limbs that have been weakened by shade.

Bushy dogwoods like cornelian cherry and Tatarian dogwood are handled like any multiple-stemmed shrub. Thin out up to one third of the old stems in the early spring to encourage new growth without spoiling the appearance of the shrub. Species that spread from underground runners may produce unwanted suckers that should be removed. Cornelian cherry is sometimes trained to a single leader and grown as a small tree. Bushy, rough-twigged species of the Tatarian dogwood grown for their red branches are often pruned heavily in early spring to produce new stems that are a brighter color than old ones.

COTONEASTER

Whether in the form of ground covers, upright shrubs or small trees, the versatile cotoneasters all have graceful spraying branches, grow rapidly and bear bright berries that ripen in fall and cling through late winter. They may be either deciduous or evergreen. Tall species should keep their natural fountain-like shape. Others lend themselves to training on walls. Prune deciduous cotoneasters in winter and evergreen species just before spring growth begins.

Tall-growing cotoneasters often require no other pruning than that provided when berry-hung branches are cut for house decoration. Such cuts should be made to preserve the plant's arching habit of growth. If necessary, thin out growth that becomes too twiggy or crowded, and remove oversized branches that throw the shrub's appearance off balance.

Thin old wood from cotoneaster ground covers to allow air to circulate among the plants. To train cotoneaster to a wall, fan out the main stems against the wall, thinning out enough branches so the wall can be seen through the plant. If greater bushiness is desired, cut back new shoots to three to five buds in winter or early spring. Remove outward-growing branches so that the plant will remain flat against the wall.

CRAB APPLE See *Malus*
CRAPE MYRTLE See *Lagerstroemia*

CRATAEGUS (hawthorn)

These thorny deciduous trees are grown either as lawn specimens or hedges and are pruned accordingly. Some hawthorns are columnar, others spreading, but all types have a tendency toward multiple trunks and dense interwoven branching. They can be trained to single trunks if desired, but they are more frequently permitted to follow their natural multiple-trunk configurations.

Hawthorns grown as hedges may be sheared throughout the summer as needed to maintain their form, but ornamental specimens should be pruned only when the trees are dormant, in winter or early spring. Thin plants to let light and air into dense crowns, and remove spreading branches that have become so long and heavy that they threaten to break in a wind or even under their own weight. Thin the trees also to remove crossing branches, but not so much as to interfere with the branches' picturesque zigzag shape. Remove suckers at any time.

CROSS VINE See *Bignonia*
CURRANT See *Ribes*

D

DAVIDIA (dove tree)

The dove tree, with its slender arching branches and huge fluttering flower bracts, is naturally multiple stemmed but is frequently trained to a single trunk, or leader. It grows to a height of 20 to 40 feet and is very dense. Blossoms do not appear until the tree is 10 to 15 years old, and then they may appear only intermittently.

Prune the dove tree in winter and preferably only when young, as it heals slowly. To train a young tree, cut back its lateral branches to encourage a central leader. Thin out some of the smaller branches to emphasize the tree's spreading shape and to expose its dark, patterned trunk.

DEUTZIA

In cold areas this popular bushy shrub, prized for its abundant flowers, generally needs pruning to remove winter damage. Low-growing species like *Deutzia gracilis* may also be pruned to keep them compact when they are used as hedge or border plants, while the taller *Deutzia lemoinei* may need to be pruned to keep it from becoming leggy.

Remove dead and broken twigs in the early spring, but for other pruning wait until the flowers fade because they bloom from buds formed the previous year. As new shoots emerge from the base of the plant, pinch them to encourage branching. Thin out the low-growing deutzias every five years, removing old nonproductive stems. Taller deutzias, however, sometimes need to be thinned more often to encourage new growth at the bases of the stems.

DIOSPYROS (persimmon)

Grown for its brilliantly colored, delectable fruit as well as its glossy leaves that turn yellow to orange in autumn, the persimmon is a highly ornamental tree. Columnar common persimmon grows 30 to 60 feet tall while the smaller Japanese persimmon is rounder and more spreading; both begin to bear fruit within four years of planting.

Train persimmon to a single trunk, and develop a strong scaffold by removing all except three to five wide-angled branches spaced about 12 inches apart; cut these branches back to about 8 inches. As the tree grows, continue cutting new side shoots back about 6 inches for every foot of new growth to encourage earlier fruit bearing. But heavy crops can break limbs and fruit itself may need to be thinned.

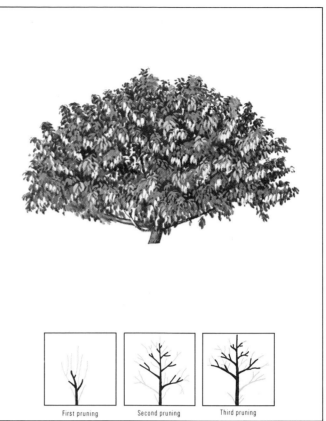

DOVE TREE
Davidia involucrata

FUZZY DEUTZIA
Deutzia scabra 'Pride of Rochester'

117

RUSSIAN OLIVE
Elaeagnus angustifolia

First pruning   Second pruning   Third pruning

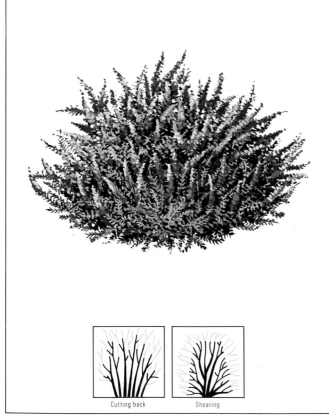

SPRING HEATH
Erica herbacea

Cutting back   Shearing

Prune mature persimmon trees in winter while they are dormant, thinning them to allow light into the tree's center. But prune lightly, to avoid stimulating water sprouts and to preserve as much new fruit-bearing growth as possible.

DOGWOOD See *Cornus*
DOUGLAS FIR See *Pseudotsuga*
DOVE TREE See *Davidia*

E

ELAEAGNUS (Russian olive)

The best-known elaeagnus is the Russian olive, but there are some 40 other kinds of this tough, resilient multiple-stemmed shrub or small tree. Most are deciduous, though a few are evergreen, and they range in habit of growth from upright to spreading. Impervious to deep cold and hurricane-force winds, they are usually grown as hedges and windbreaks, though the Russian olive is sometimes trained to a single trunk and used as a specimen tree. All the elaeagnuses are fast-growing and react to heavy pruning by sending up vigorous shoots of new wood along branches and from the base of the plant.

Train a young Russian olive tree to a single trunk in early spring by selecting and staking a strong, straight stem and removing all the others. For several years remove the lowest side branches until clearance beneath the tree is the desired height. Then let the tree branch naturally. Prune mature plants, both bush and tree forms, in late spring or early summer just after they have flowered. Cut back new shoots to half their length if necessary to thicken the plant and remove suckers. Thin out Russian olive to reveal its gnarled trunk and shredded bark. In early summer the Russian olive is prone to a disease that may require severe pruning; diseased trees may be cut back to stumps and still survive, but be sure to disinfect your pruning tools between each cut.

ELDERBERRY See *Sambucus*
ELM See *Ulmus*
ENGLISH IVY See *Hedera*

ENKIANTHUS

These tall-growing deciduous shrubs, often used as background plants in border plantings of azaleas, grow slowly and seldom need pruning to maintain their shape. They flower in spring and produce a brilliant display of red foliage in autumn. Though naturally open in their growing habit, they can be made bushier by cutting back new growth after spring flowering. To rejuvenate plants, thin out half the old stems in early spring before new growth begins.

ERICA (heath)

Although one heath reaches heights of over 18 feet, most are small, bushy plants grown chiefly as ground covers. The evergreen leaves are short and narrow, but so thick that the plants give the effect of a green carpet. Heaths bloom in early spring, late summer or winter, depending on the species, producing tiny bell-shaped flowers. To encourage flowers and new growth on hardy spring-blooming species such as *Erica carnea,* shear back plants to one half or two thirds their height immediately after blooming. A later pruning would destroy the buds, which form in midsummer and open the next spring. Do not shear tender summer- or fall-blooming species; simply snip off stalks of faded flowers. They may also be lightly pruned in early spring. On gold- or bronze-leaved species, immediately cut out shoots bearing green leaves or they will soon dominate the plant.

ERIOBOTRYA (loquat)

The subtropical evergreen loquat is a small tree or large shrub that naturally develops a round, symmetrical crown. It makes a small shade tree when trained to a single trunk, while a row of multiple-stemmed loquats can create a dense hedge. The loquat tolerates any amount of pruning to keep it smaller than its usual 30-foot height, to improve its fruit production or to espalier it.

Prune loquat in the spring after fruiting. To train a low-branching, multiple-stemmed tree, cut back all stems when the plant is about 30 inches tall. If you prefer a higher canopy and a single trunk, allow the tree to grow 6 to 8 feet high, select a leader, remove lower branches, and cut back the growing tips of the remaining side branches. Tubbed specimens look best when thinned to display the graceful branches and dramatic dark green leaves.

Loquat's fragrant white flower clusters appear in late summer or fall, followed in early spring by edible, pear-shaped fruit. For best fruit production, thin branches in spring to allow light into the tree's interior and remove half of each flower cluster to increase fruit size.

EUCALYPTUS

These fragrant, fast-growing evergreen shrubs and trees seldom need pruning, but they can be cut back severely if necessary, since they have a remarkable capacity for regrowth. The shrubby, many-stemmed species require the least amount of pruning; they naturally form thick clumps of foliage. If one of these shrubs should outgrow its allotted space or become leggy, cut it to the ground to force new shoots. Then allow three to five of the strongest shoots to form a new framework for the shrub, and remove the rest.

Medium-sized eucalyptus trees can be pruned to develop one or many trunks. To encourage a single trunk, remove side shoots as soon as the tree is established. To encourage several trunks, allow three or four of these side shoots to continue to grow, and cut back the main trunk to a height of 2 feet. If side shoots fail to appear, bend the trunk to the ground and hold it in an arched position by wiring it to a stake. Within a year, new shoots will sprout from the base of the old trunk, which can then be removed.

Tall single-trunk eucalyptus trees need maintenance pruning from time to time to remove damaged or weak wood. Prune in spring or summer after flowers fade by cutting just above a side branch or bud.

EUGENIA (rose apple, Surinam cherry, Australian brush cherry)

Fruit-bearing tropical evergreens, the eugenias are popular in mild climates as subjects for topiary and espalier work and are also planted in hedges. They have dense foliage and are very tolerant of heavy and frequent pruning. This can be done at any time, though the preferred time is the beginning of the growing season. Later pruning will reduce the production of aromatic flowers and spicy, edible fruits.

Eugenias can be trained to grow as trees if a central leader is allowed to dominate and competing leaders are shortened. Once such a tree becomes about 10 feet tall, little further pruning is needed.

EUONYMUS (spindle tree, strawberry bush, wintercreeper)

Grown for colorful foliage and autumn berries, the various euonymus species are trained as shrubs, hedges, ground covers, vines and small decorative trees. In mild climates these hardy plants can be pruned at almost any time, but in colder areas prune in the spring after danger of frost is past.

First pruning · Second pruning · Thinning · Espaliering

LOQUAT
Eriobotrya japonica

First pruning · Second pruning · Third pruning

RED FLOWERING GUM
Eucalyptus ficifolia

DWARF WINGED EUONYMUS
Euonymus alata compacta

Thinning / Cutting back / Shearing

POINSETTIA
Euphorbia pulcherrima

Cutting back / Rejuvenating / Pinching

For deciduous types grown as shrubs, such as winged euonymus, spindle tree and strawberry bush, thin out about one third of the old growth each year and cut back the tips of new growth to keep the plants shapely.

Evergreen species such as wintercreeper can be trained as ground covers by allowing them to spread and by keeping their height under control; shear off upright shoots and the tops of mounds. To grow them as vines, train them to supports and cut back long shoots so plants will lie flat against the supports. When grown as shrubs, the evergreen species should be pruned to remove deadwood and may be cut back or sheared to maintain shape. To grow them as hedges, cut back plants to 6 inches when they are planted, and continue cutting back 6 inches for every foot of growth until the desired height and shape are attained; make sure the bottom is wider than the top. A tree-form euonymus is trained by cutting off the lower limbs and encouraging a strong central leader; pinch back the top growth to shape the tree.

EUPHORBIA (poinsettia)

The Christmas-blooming poinsettia bears its brightly colored flower-like bracts at the tips of new branches. Generally grown as an indoor plant 12 to 30 inches high, in Florida and Southern California it is an outdoor plant, reaching a height of 10 feet or more, and is treated as a shrub.

For Christmas color, cut back poinsettia stems to three or four leaves in spring, then pinch back new growth to encourage branching until about mid-August. For fewer but larger bracts, remove small branches and weaker stems during the spring. Rejuvenate old plants in the spring by cutting them back to about 12 inches above the ground, forcing the growth of new shoots. When these shoots have developed four or five leaves, prune them as above.

EXOCHORDA (pearlbush, Wilson pearlbush, common pearlbush)

Stringlike clusters of round white flowers, borne in the spring, give this shrub its common name. The shrubs grow up to 15 feet tall. To counteract a natural tendency toward gangliness, pinch off the tips of young shoots from time to time after flowering; this will channel energy into new side shoots. Weak branches should be removed, since only the strong ones produce an abundance of flowers, and flowers are the main reason for growing the shrub. Such pruning should be done immediately after the flowers fade. Every three or four years, remove any suckers that have developed at the base of plant and thin some of the oldest and weakest wood down to the ground.

F

FAGUS (beech)

All forms of beech—columnar, pyramidal or weeping—grow 50 to 90 feet tall and almost as wide. Unpruned, these deciduous trees begin to branch at ground level, a growth habit that makes it possible to shear beeches into hedges.

Young beech trees should be trained to a single trunk. Remove lower branches to a comfortable height if headroom is desired. Prune to correct narrow crotches, as beech wood is brittle and splits easily. Mature trees need only maintenance pruning to remove dead, diseased or crowded branches.

To develop a hedge, cut newly planted young beeches back to 1 to 2 feet to encourage side growth. Shear them annually in late winter or early spring or when new growth hardens in late summer. Beeches are usually pruned into hedges 10 to 12 feet high and with several shearings yearly can be restricted to widths as narrow as 1½ feet.

FALSE CYPRESS See *Chamaecyparis*

FATSIA

Multiple-stemmed, evergreen fatsia is grown mainly for its large, deeply indented leaves, although the plant bears clusters of small, white flowers in the fall and early winter; the flowers are followed by blue berries. Fatsia is fast growing and tends to become lanky and bare-stemmed; it also sends up suckers. In early spring, thin out bare stems, cutting them off about 6 inches from the ground; this forces new shoots from below. Allow several of the strongest shoots to develop into a new crop of stems and remove the others. Fatsia can be grown as a small single-stemmed tree by removing all but one stem and pruning off its lower branches. Remove flower heads as they form to promote larger leaves. Old fatsia plants can be rejuvenated by cutting off all their stems to about 6 inches above the ground in spring.

FERN PODOCARPUS See *Podocarpus*

FICUS (fig)

Fig trees are deciduous and frequently bear two crops of edible fruit a year, the first in early summer on the previous year's growth, the second in late summer on the current season's growth. The trees grow naturally into round, spreading specimens reaching up to 40 feet tall. As ornamental plants, they need only maintenance pruning to remove deadwood or crowded branches. Any heavy pruning that is necessary should be done in winter to avoid interference with the production of fruit.

To train a fig tree primarily for fruit production, remove all but three or four wide-angled branches along the trunk of the young tree, spacing them about 12 inches apart, and cut back these branches to 6 or 8 inches. Thereafter, thin and cut back lightly in winter to keep trees well-shaped and to open their centers to light. In the garden, fig trees will bear fruit about three years after they are planted; trees that are grown in tubs will bear sooner.

FIG See *Ficus*
FIR See *Abies*
FIRETHORN See *Pyracantha*
FLAME BUSH See *Calliandra*
FLAME-OF-THE-WOODS See *Ixora*
FLOWERING QUINCE See *Chaenomeles*

FORSYTHIA

Naturally upright or arching multiple-stemmed shrubs, forsythias flower in early spring from buds formed during the previous year. When the plants are five to six years old, they will probably need to be rejuvenated by thinning out the old wood, which is darker and has a coarser texture than the younger growth.

To thin a mature shrub, remove one third to one fourth of the old canes each year immediately after flowering. Alternatively, prune the plant four to six weeks before flowering and take the cut branches indoors to force the buds into early bloom. If a forsythia plant becomes so crowded that it cannot be thinned effectively or is badly shaped, cut the whole plant to about 1 foot from the ground. Then remove all the heaviest canes at ground level, leaving only half a dozen younger canes. Do this soon after flowering to allow maximum time for new growth to become established before frost. When new growth is about 2 feet high, pinch off the terminal buds to encourage branching. The shrub will take about two years to develop a good shape again.

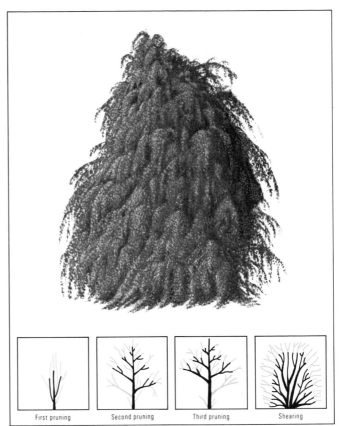

WEEPING BEECH
Fagus sylvatica pendula

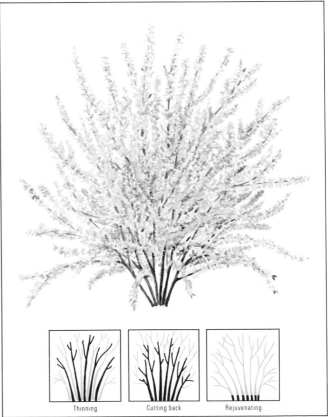

BORDER FORSYTHIA
Forsythia intermedia 'Beatrix Farrand'

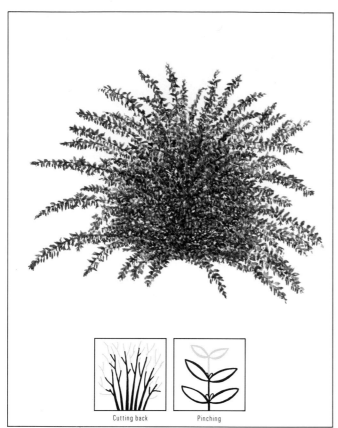

MAGELLAN FUCHSIA
Fuchsia magellanica

GARDENIA
Gardenia jasminoides

FRAXINUS (ash)

These hardy deciduous trees grow quickly to heights of from 20 to 100 feet or more depending on the species. Prune young ash trees severely in early spring to develop a central leader, or trunk, and strong scaffold branches. Well-spaced branches with wide crotches are particularly important because ash wood is brittle and liable to damage from wind or the weight of snow. If they are shaped properly when young, ash trees need little pruning as they mature. In early spring, remove deadwood and the water sprouts that tend to sprout vertically from the branches. In the winter, cut back weak branches or those that lean out of the tree at an awkward angle to the nearest strong crotch.

FRINGE TREE See *Chionanthus*

FUCHSIA

A semitropical shrub, the colorful fuchsia comes in two forms, trailing and upright, both of which can be heavily pruned to create special effects. Fuchsias can be trained to the shape of small trees, or standards. Most fuchsia species are grown outdoors only on the West Coast and in the Southwest; elsewhere they are popular as greenhouse or container-grown terrace plants.

Fuchsias, whether trailing or upright, are generally trained to bushy crowns by pinching back new growth. To promote more blooms in the following year, cut back branches as soon as the plant has flowered so new shoots will have time to develop during the current growing season.

G

GARDENIA

In frost-free areas, gardenias are evergreen garden shrubs, growing to a height of 3 to 6 feet, but in the north they are raised indoors, and are pruned to limit their height to about 30 inches. This encourages them to produce more blossoms—and the white, waxy, exceptionally fragrant flowers of gardenias are one of their chief attractions.

Gardenias bloom from late spring to summer outdoors, and in winter when grown indoors. They produce their most spectacular flowers in their third, fourth and fifth years; after that they may become leggy, with smaller buds and leaves, although careful pruning can keep the plants in their prime.

Prune gardenia plants immediately after they have flowered. Remove faded blossoms and snip off the stem tips to force new growth. Thin to remove overcrowded branches and those that are producing weak blossoms or abnormally small leaves. Shape the plants by cutting back overgrown branches. Rejuvenate a leggy indoor gardenia by cutting back all stems to 6 inches above soil level, thus forcing new growth from the plant's base. To produce fewer but larger flowers, pinch off all but a few buds on each stem.

GINKGO (maidenhair tree)

Time, rather than pruning, will correct the awkward appearance of an immature ginkgo. As ginkgo trees grow, their trunks may divide into upright branches that will gradually spread apart over some 20 years to form magnificent open crowns. These deciduous trees eventually reach 50- to 80-foot heights but may grow only 8 to 12 inches annually. Ginkgoes heal slowly, so pruning should be kept to a minimum. Stake a young ginkgo tree and prune it to establish a strong central leader and good scaffold branches. Then confine pruning to winter or early spring removal of side branches that grow out from the upright scaffold branches at bizarre angles. Otherwise they will need no pruning.

GLEDITSIA (honey locust)

Thornless varieties of the deciduous honey locust develop rapidly into 35- to 70-foot shade trees with short trunks and spreading, open crowns.

Stake young honey locust trees and prune them to develop a central leader, or trunk, and strong scaffold branches. Cut the side branches back severely at the time of planting. As the trees grow, remove any branches that grow from the scaffold branches at less than a 45° angle; some varieties of the honey locust have a tendency to form weak crotches. When the fork between two branches is too narrow, cut one of them back to a side branch or bud so that the remaining branch will be dominant.

To train young honey locusts for hedges, cut the trees back to 1 foot at the time of planting. Shorten side shoots to 10 inches as they develop, and pinch the tips of new growth to encourage bushiness until the hedge is 8 to 10 feet tall.

GLORY BUSH See *Tibouchina*
GOLDEN CHAIN See *Laburnum*
GOLDEN RAIN TREE See *Koelreuteria*
GOOSEBERRY See *Ribes*
GRAPE See *Vitis*
GRAPEFRUIT See *Citrus*
GREEN EBONY See *Jacaranda*

H

HACKBERRY See *Celtis*

HAMAMELIS (witch hazel)

Witch hazel, source of a fragrant astringent lotion, can be grown as a multiple-stemmed shrub or a small tree. As a shrub it seldom needs pruning except to open a crowded plant or rejuvenate one that has passed its prime. To open a plant, thin out some of the side branches; to rejuvenate it, thin out some of the older stems to ground level. Witch hazel shrubs may also be cut back to limit their size. The correct time for pruning depends on the season of flowering. Witch hazels that bloom in the spring are pruned after flowering, and fall-flowering varieties are pruned before new growth appears in the spring.

To train witch hazel as a tree, choose one stem as a central leader, or trunk, stake it and remove all the other stems. Each year, for several years, remove the side branches and the suckers that form at the base of the plant.

HAWTHORN See *Crataegus*
HEATH See *Erica*
HEATHER See *Calluna*
HEAVENLY BAMBOO See *Nandina*

HEDERA (English ivy)

When it is grown against walls or used as a sprawling ground cover, English ivy is such a vigorous plant that it can be pruned at almost any time of the year. However, you should avoid heavy pruning in fall or winter unless you are willing to wait for new spring growth to fill in the resulting bare spots. Thin climbing ivy regularly to lighten its weight; old vines that have become heavy can fall away from walls, pull down gutters and damage fences and house siding. Cut back straggly shoots to keep vines flat against walls. Old English ivy vines can be rejuvenated by cutting them back severely in early spring.

HEDGE THORN See *Carissa*
HEMLOCK See *Tsuga*

MORAINE THORNLESS HONEY LOCUST
Gleditsia triacanthos inermis 'Moraine'

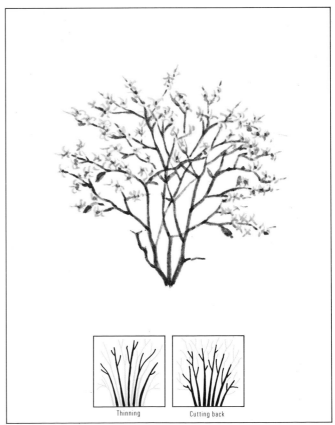

CHINESE WITCH HAZEL
Hamamelis mollis

CHINESE HIBISCUS
Hibiscus rosa-sinensis

BIG-LEAVED HYDRANGEA
Hydrangea macrophylla

HIBISCUS (rose of Sharon, shrub althea)

Shrub-type hibiscus like the rose of Sharon is pruned to control its size. Left to grow naturally, it spreads with age and reaches a height of 12 to 15 feet. Cut it back in winter or early spring to remove crowded branches. For larger flowers, shrub hibiscus can also be cut back in spring, leaving only two buds of the previous year's growth on each branch.

Rose of Sharon is sometimes pruned into an informal hedge by cutting back side growth to encourage upward growth, but the plant should never be sheared, as this interferes with its production of flowers. It can also be trained into the shape of a small tree, or standard, by selecting one stem as a trunk and removing all the others. Cut back the side branches for the first year to encourage the leader, and remove suckers that may sprout from around the base. When the tree is established, prune only to maintain its shape.

HILLS-OF-SNOW HYDRANGEA See *Hydrangea*
HOLLY See *Ilex*
HOLLY GRAPE See *Mahonia*
HONEY LOCUST See *Gleditsia*
HONEYSUCKLE See *Lonicera*
HONG KONG ORCHID TREE See *Bauhinia*
HOP HORNBEAM See *Ostrya*
HORNBEAM See *Carpinus*
HORSE CHESTNUT See *Aesculus*
HORTENSIA See *Hydrangea*

HYDRANGEA (bigleaf hydrangea, hills-of-snow hydrangea, hortensia, peegee hydrangea)

Hydrangeas are pruned according to type and function. The summer-blooming species such as hills-of-snow are often used in shrub borders or as informal hedges to mark a property line. The summer-to-fall-blooming peegee hydrangea, however, may reach a height of 20 feet or more and is often trained to a single-trunked small tree or formal standard. Both plants flower on the current season's growth and are pruned in early spring before new growth begins.

For a profusion of medium-sized blossoms borne on strong stems, cut back hills-of-snow by half; more severe pruning may create larger-sized flowers, but they will be too heavy for the stems to support.

To train peegee hydrangea to a single trunk, remove all stems but one, and cut off branches until the trunk is the desired height. To control the size of the plant and produce larger flowers, cut back branches in the spring, leaving only two buds; then thin out new shoots, removing the weaker ones when they are about 10 inches long.

Hortensia, which is a tender spring-blooming species sometimes grown as a house plant, is used in the garden as a specimen plant or border shrub, where it reaches a height of 3 to 6 feet. Hortensia blooms on buds set in the previous season, so it should be pruned right after flowering. Cut it back only to control its size and to remove dead branches.

I

ILEX (holly)

The most spectacular species of these handsome, mostly evergreen shrubs and trees are prized for their red winter berries and glossy, spiny-edged leaves. Although they require little pruning except when young, to train them, and later to remove dead, diseased or damaged wood, hollies can in fact be cut back severely and are sometimes used for sheared hedges or topiary. The plants are generally pruned in late winter or early spring, or around Christmas when branches can be used for wreaths and table decorations.

In shaping a holly, cut back only to a side branch or bud; if a branch is cut back to the trunk, new growth may never appear to replace it. Remove any branches that droop to the ground, for these may take root and create unwanted growth. To train a young plant to a central leader, remove competing stems. If the leader is damaged, cut back to healthy wood just above a bud or leaf; if the leader is dead, cut back to the nearest whorl of branches and train one of these branches to become the new leader.

INDIAN BEAN See *Catalpa*
IRONWOOD See *Carpinus*

IXORA (jungle geranium, flame-of-the-woods)

A spectacular flowering shrub for use in hedges and foundation plantings in subtropical climates, the ixora bears 4- to 6-inch clusters of brilliantly colored flowers almost continuously. In other regions ixoras are often grown in containers in the greenhouse.

Pruning can be done at any time of the year. Plants with a loose, natural, globular shape bear the most flowers, but ixoras can be trained as formal hedges. To do so, cut the young plants back to 6 inches after planting, then continue cutting back about 6 inches for every foot of growth until the desired height is reached. The plants can be expected to reach 5 feet in about five years and may ultimately become 15 feet tall. In hedge pruning, make sure the bottom is wider than the top, or lower branches are likely to die.

Specimen plants in the garden or ixoras grown in containers can be prevented from becoming overcrowded by thinning out the old stems in the early spring. If a plant develops a poor shape, it can be cut back to a height of 8 to 12 inches. Pinching young shoots when they are about 8 inches long will encourage branching.

J

JACARANDA (green ebony, sharp-leaved jacaranda)

Naturally open and spreading, the jacaranda is a warm-climate flowering tree that requires little maintenance pruning once it has been trained to grow with either a single leader or multiple trunks. It quickly reaches a height of 25 feet and at maturity may be 50 feet tall. In late spring and early summer, the jacaranda bears 8-inch-long clusters of fragrant blue flowers; when the blossoms fall, they carpet the ground beneath the tree. Limbs that are bent near the breaking point by the weight of their flowers and foliage should be thinned by removing some of the weaker branches back to the nearest crotches.

Because the jacaranda is fast-growing, its branches are brittle and subject to wind damage; regular maintenance includes removal of such broken limbs as well as small dead branches to be found near the center of the tree.

A young single-stemmed tree can be trained to a multiple-stemmed plant by cutting it back to a height of 2 feet and allowing several stems to develop. Remove growing tips for every 3 feet of growth to encourage branching. Trees badly damaged by frost can sometimes be saved by cutting back to live wood and letting one or more new shoots develop.

JAPANESE PAGODA TREE See *Sophora*
JASMINE See *Jasminum*

JASMINUM (jasmine)

Jasmine plants grow mostly in warm climates and they can be trained as either shrubs or vines (though they must be attached to walls or trellises, as they will not climb unaided).

AMERICAN HOLLY
Ilex opaca

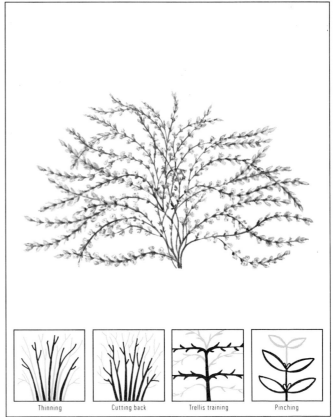

WINTER JASMINE
Jasminum nudiflorum

BLACK WALNUT
Juglans nigra

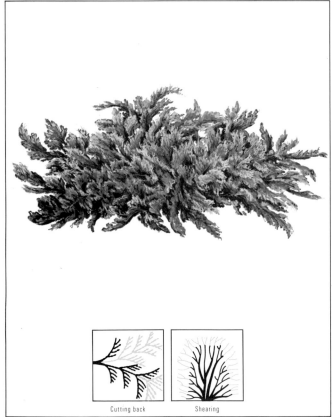

WILTON CARPET JUNIPER
Juniperus horizontalis wiltonii

Most are evergreen and generally need little pruning beyond thinning out old canes and cutting back overly long ones. One exception is the winter jasmine, which may, in the South, reach heights of 15 feet or more if it is not pruned, although farther north it usually remains a manageable 3 to 8 feet in height.

All jasmines blossom on the previous season's growth, and should be pruned immediately after flowering so as not to interfere with next year's bud formation. But the various species blossom at different times of year: winter jasmine in late winter and early spring, Italian jasmine in midsummer to fall, primrose jasmine in fall to spring. To encourage a jasmine plant to branch laterally in the form of a shrub, pinch back the tips of upright canes; then shape lateral branches by cutting them back to half their length.

JUGLANS (walnut)

Excellent as shade trees, the American black walnut and the smaller English walnut are also prized for the harvest of nuts they provide in the fall after they are four to eight years old. Both trees are fast growing: within six to eight years of planting, they can be 20 feet tall, well on their way to mature heights of 50 feet or more.

Walnuts are usually pruned in summer or fall, since they bleed heavily if cut in spring. Train young trees to a single leader, or trunk, and select three to five sturdy branches for the scaffold. These branches should be spaced about 2 feet apart and begin 5 to 6 feet from the ground. Shorten these branches by half. In subsequent years thin as needed to remove suckers, water sprouts and low-growing, damaged, crowded or crossing branches. Mature specimens need little maintenance pruning except removal of deadwood.

JUNGLE GERANIUM See *Ixora*
JUNIPER See *Juniperus*

JUNIPERUS (juniper, Colorado red cedar, eastern red cedar)

Junipers are easily pruned to control their size and shape. However, because these evergreens come in a wide variety of forms—tree, upright shrub, twisted, prostrate and ground cover—gardeners should match each plant to a specific landscaping purpose. The juniper looks best when its natural outline is accentuated by light pruning. Cut back individual branches to branchlets in early spring before new growth begins. To control a plant's size, cut back up to half of the new growth in summer and again in fall if needed. Some junipers can be sheared and used as hedges, but when they are used as hedge plants they are less amenable to heavy shearing than such evergreens as yews or arborvitae because junipers will not produce new shoots as readily when they are cut back to old wood.

Tree junipers such as the eastern red cedar, Rocky Mountain juniper and Colorado red cedar need little pruning. On young trees with competing leaders, select the strongest leader and remove the others. An errant branch may be cut back to a fork. Thin out upper branches that block light and air from lower ones by cutting them back to another side branch or to the trunk.

Prune plants to enhance the natural shapes of shrubs and ground covers. Ground cover and prostrate types such as the Gold Coast, Sargent and shore junipers may be kept low by cutting back vertical branches in early summer. Compact conical types or hedge types such as the California juniper can be lightly sheared in the summer. Control spreading varieties such as the Pfitzer juniper by cutting back straggly horizontal branches and removing undesirable vertical ones before new growth begins in the spring or when needed.

To rejuvenate bushy types, cut back straggly branches as far as necessary while the plants are dormant. Shorten remaining branches by cutting each back to the first branchlet.

K

KALMIA (mountain laurel)

Mountain laurels, like their relatives, the azaleas and rhododendrons, need little pruning to maintain their graceful shape. These hardy evergreen shrubs can grow 4 to 8 feet tall in 10 years, but they can be kept to a smaller size without much difficulty.

Prune, if necessary, after plants flower in late spring. Removing clusters of faded blossoms is beneficial to next year's flower growth, though not essential. Prune badly damaged branches back to the nearest fork of healthy wood and cut back overgrown shoots if space is a problem. Straggly or leggy plants may be reshaped by cutting old, overgrown stems to ground level before new growth begins; this encourages new growth from the base. For shrubs that are badly deteriorated, remove all the stems at ground level; the entire plant will be rejuvenated in two to three years.

KATSURA TREE See *Cercidiphyllum*
KIWI BERRY See *Actinidia*

KOELREUTERIA (golden rain tree, Chinese flame tree)

These ornamental deciduous trees with clusters of yellow flowers grow 20 to 40 feet high and have naturally open crowns that require little pruning to maintain their shape. In fact they are often grown for their wintertime appearance, with twisted branches and puffy bright orange or brown seed pods that look particularly decorative when covered with snow. The Chinese flame tree is subject to winter damage in severe cold. Spring- and summer-flowering species are pruned similarly. In winter, thin out crossed or crowded branches; in spring, remove shoots that appear on the trunk below the point where it divides into the main branches, as well as all shoots of twiggy growth that appear at the crotches. Thin out dead branches.

KOLKWITZIA (beauty bush)

This graceful, arching shrub should not be pruned for either shape or size. It should be allowed to grow to its natural height of from 6 to 12 feet and to spread fountain-like from a tight group of canes to a breadth of 15 feet or more. If given sufficient room, it requires only an annual thinning of a few of its oldest canes and deadwood to keep it healthy. Remove these at ground level or at the nearest new branch after the bush has flowered, taking care not to damage adjacent canes growing close beside them.

L

LABURNUM (golden chain)

Laburnum is not a sturdy tree—its roots are weak and cuts heal slowly—but its spectacular spring display of long golden flower clusters justifies the extra care it requires. It should be pruned during the early summer to remove seed pods and dead branches, and to encourage the formation of new flowering shoots for the following year. The tree should be trained when young. It naturally develops a single leader or several trunks, and the branches angle out sharply, producing a vase-shaped crown. This graceful shape should be encouraged by thinning out any shoots that tend to grow horizontally. Every part of the laburnum is highly toxic; do not allow any part of it to come in contact with your mouth.

MOUNTAIN LAUREL
Kalmia latifolia

WATERER LABURNUM
Laburnum watereri

CRAPE MYRTLE
Lagerstroemia indica

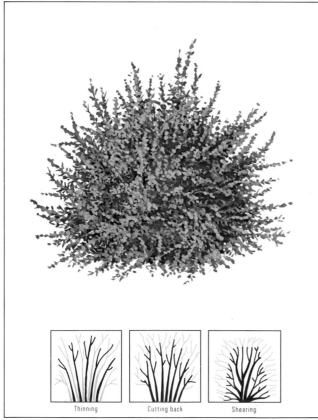

NEW ZEALAND TEA TREE
Leptospermum scoparium hybrid

LAGERSTROEMIA (crape myrtle)

This summer-flowering shrub blooms at the tips of the current year's growth; it should be pruned in late winter or early spring while it is still dormant. Some species grow as much as 20 feet tall and may need pruning to control their size, but there are smaller types, only 5 to 7 feet high, for which this kind of pruning is seldom necessary. Larger crape myrtles are used as specimen plants and can be trained to single stems, as decorative roundheaded trees. Smaller species are used as accent plants in gardens and in tubs and are ideal for informal hedges.

In early spring, remove any winter-damaged branches and cut back the ends of branches to increase bushiness. Thin out some of the new stems annually so that the base of the plant does not become too crowded. Every five or six years, thin out some of the older stems to encourage new growth. Crape myrtle bushes are sometimes cut back to the ground completely when they become oversized or when they suffer severe winterkill; the flowers on plants that are pruned in this manner will be enormous.

To train a crape myrtle to a single trunk, cut away all but one strong stem of a young plant, stake it, and remove all suckers and new shoots that form along the trunk until it is 6 feet tall. Pinch back top growth lightly during the growing season until the head of the shrub becomes dense and rounded. Thereafter cut back lightly each year after flowering to keep the head shapely.

LANTANA (common lantana, trailing lantana)

In frost-free areas, trailing lantana is used as a ground cover that provides a carpet of 1-inch clusters of tiny lavender flowers; it can be kept under 18 inches tall by pinching back new growth, and it can be kept vigorous by cutting out old, woody stems and any dead undergrowth.

Elsewhere, both trailing and upright lantanas are grown in containers and moved indoors for the winter or are treated as annuals. Since lantana responds to severe pruning by producing lush new growth, the common variety is often grown as a standard with a plump head of foliage and flowers atop a 2- or 3-foot stem.

If the tops of outdoor plants are damaged by frost, wait until spring to remove deadwood. The plants can be cut all the way back to a few inches aboveground if necessary. Lanky growth can be cut back at any time. In cold climates, outdoor plants can be cut back severely, potted and moved indoors several weeks before frost.

LARCH See *Larix*

LARIX (larch, tamarack)

Although they lose their needles in the winter, larch trees are cone-bearing, like pine and spruce. Fast-growing young trees develop narrow pyramid shapes that gradually broaden with age. Their wide horizontal branches rarely need pruning, but they can be trimmed for a symmetrical appearance. Mature trees bear showy red-purple male flowers in the early spring; these pollinate the green female flowers that become long-lasting upright cones. It is safe to prune larches lightly at any time.

If the leader of a young tree breaks or is damaged, replace it by tying the side shoot nearest the top in an upright position. Thin out broken branches from old trees. Remove low branches for the desired amount of headroom.

LAUREL See *Laurus*
LAUREL, MOUNTAIN See *Kalmia*

LAURUS (laurel, sweet bay)

In mild climates where the evergreen laurel will flourish, it is a favorite choice for use in topiary sculpture because it tolerates hard and frequent pruning so well. It can be pruned selectively in spring or summer, or it can be sheared when new growth is 3 or 4 inches long. Cut ends and damaged leaves are rapidly hidden by new growth. Tub-grown specimens, too, can be shaped into cones, pyramids, globes and stately standards, and the plant lends itself well to the frequent shearing needed to maintain it in a formal hedge. Cut branches may be used for long-lasting indoor decorations.

Left to grow naturally, laurel produces a slender, conical head of dense, aromatic foliage on multiple stems.

LEMON See *Citrus*

LEPTOSPERMUM (tea tree)

These flowering semitropical evergreen plants are used as informal hedges or small trees, depending on the species. All of them flower profusely, the flowers appearing on growth of the previous year. Major pruning should be done immediately after flowering but light shaping may be done at any time.

Smaller species, like the New Zealand tea tree, are naturally compact and require little pruning. They can be clipped to form a hedge by removing some of the side branches, but do not cut them back beyond existing foliage because new growth will not develop on sections of branches that are no longer bearing leaves. To keep mature plants from becoming crowded, thin out a third of the oldest stems each year.

The larger Australian tea tree is so sprawling and spreading that it must be pruned severely when young to establish a controlled shape. It too can be clipped as a hedge, but it is often grown as a specimen tree with graceful drooping branches; as such it will reach a height of 20 or 30 feet. Train a young tree to a single trunk or to several trunks by removing unwanted stems and lower branches. Cut back the remaining branches until the trunk is established; then prune annually to maintain the desired shape.

LESPEDEZA (bush clover)

Left unpruned, this fast-growing deciduous shrub reaches a height of 6 to 10 feet. But at that height it tends to look unkempt and so it is seldom permitted to grow more than 3 or 4 feet tall. Bush clover flowers in late summer and early fall on new growth, and should be pruned in late winter or early spring while the plant is still dormant. Thin out deadwood and crowded stems; cut back branches that are overlong, but take care not to spoil the plant's natural shape, which in some species is extremely graceful. To control the size of bush clover, or in case of winterkill, prune the entire plant back to the ground in early spring. New stems grow quickly and by late summer the plant will be 3 feet tall.

LEUCOTHOË

The most useful of these dense flowering shrubs are the evergreen species that are frequently used in foundation plantings and as cover-ups for the unsightly bases of trees and taller shrubs. The plant's long, arching stems should be pinched when young to encourage bushiness and in maturity may be cut back to keep the plant from becoming straggly. To encourage new growth, remove a few old canes every year in late February, before the growing season. To rejuvenate a failing or untidy leucothoë, cut back the entire plant to ground level in early spring. Although the flowers will be lost in the current year, new shoots will have time to harden before the arrival of cold weather.

SHRUB BUSH CLOVER
Lespedeza bicolor

DROOPING LEUCOTHOË
Leucothoë fontanesiana

SUWANNEE RIVER PRIVET
Ligustrum japonicum 'Suwannee River'

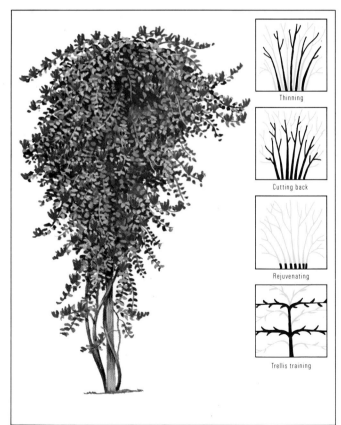

TRUMPET HONEYSUCKLE
Lonicera sempervirens

LIGUSTRUM (privet)

Privet is most commonly chosen for hedges that are to be formally trimmed because the vigorous shrubs have dense foliage and are amenable to heavy pruning and shearing. The evergreen species serve as graceful specimen or foundation plants that bloom profusely when pruned only lightly to maintain their natural shapes. Privets may be pruned to have only a single trunk and adapt easily to topiary training. In mild climates, they may be pruned year around; in cold areas they should be pruned in early spring and again in summer, as needed.

Hedges should be narrower at the top than at the bottom to allow light to reach lower branches. Shear formal hedges and topiary in early spring and repeat whenever new growth is 3 to 4 inches long. More frequent clipping during the first few growing seasons will assure a dense, compact formation.

On unsheared specimen plants, cut back a few of the oldest stems to the ground each winter to encourage new growth. Occasional tidying of the plant may be done at any time but too much cutting will interfere with bud formation. Rejuvenate failing plants by cutting the entire plant back to 6 to 12 inches above the ground in early spring.

LILAC See *Syringa*
LILAC, SUMMER See *Buddleia*
LIME See *Citrus*
LIME, ORNAMENTAL See *Tilia*
LINDEN See *Tilia*

LINDERA (spicebush)

The aromatic spicebush, a multiple-stemmed deciduous shrub, usually reaches a height of 6 to 8 feet and can maintain its neat inverted-triangle shape without pruning. It is an excellent plant in shrub borders. In early spring, tiny yellow flowers appear in clusters on wood of the previous season and if male spicebushes are nearby, female plants produce scarlet berries in the fall. Any thinning or shaping should be done in spring after the shrubs have flowered. To rejuvenate old overgrown bushes, cut them to the ground to force new growth from the base; this can be done all at once or a few of the oldest canes can be cut each year.

LIQUIDAMBAR (sweet gum)

Although scraggly when young, sweet gum develops into a neat shade tree, its trunk branching high with a spread that is about two thirds of its 60- to 70-foot height. Prune sweet gum during the winter to avoid excessive bleeding. On young trees thin out any branches that threaten the dominant leader. As trees mature, thin out closely spaced branches while the branches are still small. Watch newly planted trees carefully; if any long branches threaten to throw off the symmetry of the tree, remove a few each year until the shape is made uniform. Trees can be made more compact by cutting back the ends of new shoots.

LIRIODENDRON (tulip tree, tulip poplar, yellow poplar)

The tulip tree, a member of the magnolia family, grows rapidly to a height of more than 100 feet. In a large garden it makes an excellent shade tree that seldom needs pruning; it has a naturally straight trunk and wide-spreading branches. However, the wood is brittle and subject to winter damage.

Remove unwanted growth on the trunk and branches as soon as it appears. During the late winter or early spring, remove dead, crowded and crossing branches, but do not try to prune a tulip tree yourself after it is 20 or 30 feet tall; for removal of high branches, call an expert.

130

LOCUST See *Robinia*

LONICERA (honeysuckle)

The pruning requirements of honeysuckle plants vary with the species and their uses. In general, the vine types need more attention than the shrubs, which have a tendency to be naturally neat if they are given sufficient room to grow. Most species flower and fruit on new growth, and should therefore be pruned in the winter. The evergreen winter honeysuckle, however, is an exception; it bears its flowers on the previous year's growth and should be pruned in late spring or after it has flowered.

Thin out old wood from shrub honeysuckles to encourage new growth. Remove dead branches. If branches become too woody, cut them back to the main stem. Small and medium-sized shrub honeysuckles, like Clavey's dwarf and box honeysuckle, may be grown as formal hedges, and clipped or sheared accordingly.

Some honeysuckle vines that are grown as ground cover are so vigorous that they may need to be cut back severely in early spring; otherwise they may become invasive and almost unstoppable. On other honeysuckles of more moderate growth, thin out old wood and cut back new wood enough to prevent a leggy appearance.

LOQUAT See *Eriobotrya*

M
MADRONA See *Arbutus*

MAGNOLIA

Depending on the species and the landscaping purpose, magnolias are treated as shrubs or trees, and some may be espaliered. They should be pruned in early summer for best healing. To train magnolias as shrubs, cut back the stems of young plants to encourage bushiness. To train them as trees, select a central leader, or trunk, and cut back side shoots to strengthen the leader. To espalier a magnolia, choose a pattern with widely spaced branches to encourage the best flower production.

Once trained, magnolias should be pruned only for corrective reasons; they heal slowly, and open cuts are subject to disease. Cut back damaged wood to healthy wood; remove suckers that form around the bases of some deciduous species. Pinch off spent flowers before they form fruit, since fruit production saps a plant's vigor.

MAHONIA (holly grape)

The evergreen holly grapes include upright shrubs, which are useful as foundation plants or informal hedges, and low-growing creepers, which can serve as ground covers. The shrubs require only occasional corrective and renewal pruning, while the ground covers need regular shearing to keep them dense and compact.

Prune holly grape plants in early spring, before active growth resumes. To keep shrub forms compact and upright, cut them back hard for the first few years to establish dense growth. Keep mature plants within bounds by thinning out suckers that form at the bases. Remove dead or damaged canes as soon as they appear.

Prune low-growing varieties of holly grape annually when they are young to thicken the clumps, cutting them back severely. Mature plants may be kept dense by cutting them back nearly to the ground when they become leggy.

MAIDENHAIR TREE See *Ginkgo*

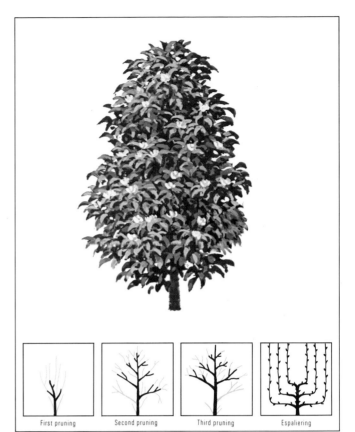

SOUTHERN MAGNOLIA
Magnolia grandiflora

First pruning Second pruning Third pruning Espaliering

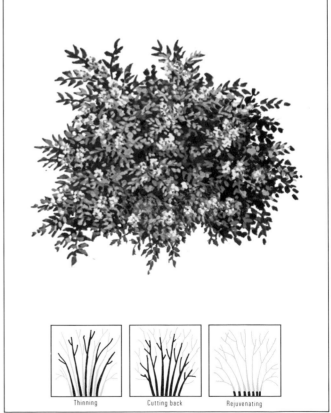

OREGON HOLLY GRAPE
Mahonia aquifolium

Thinning Cutting back Rejuvenating

MALUS (apple, crab apple)

Admired for their frothy pink and white blossoms, apples and crab apples are often planted as ornamental trees as well as for their fruit. This is particularly true of crab apple trees, whose fruit is generally under 2 inches in diameter and therefore a less desirable crop fruit than apples.

When training a crab apple or apple tree as an ornamental, select one stem as a central leader and remove any competing stems. As the tree grows, remove side branches up to about 3 feet above the ground and thin out crowded, crossed or damaged branches.

An apple tree grown for fruit production may be standard, semidwarf or dwarf in size, but all sizes are pruned in basically the same way. Plant the tree in the fall or early spring and prune the slender whip back to about 3½ feet. The following spring, select a central leader, or trunk, and three to four well-spaced lateral branches that form angles of about 45 degrees with the leader; prune these branches back to 6 to 8 inches. The next spring, cut back the tips of both the leader and the laterals to encourage the growth of more side shoots. Thereafter, prune while the trees are dormant to thin out dead, damaged, weak or crowded branches and branches that no longer bear fruit.

When pruning apples and crab apples, bear in mind that their fruit grows on short stubby shoots, or spurs. These first appear on 3- to 5-year-old trees and remain productive for up to 20 years. In spring, each spur puts forth a small amount of new growth for the current year's production and forms a fruiting bud at its tip which will develop to bear the following year's apples.

When the trees begin bearing fruit, some of the fruit should be thinned out annually to improve the quality of the crop. Apple trees thin their fruit naturally in a phenomenon known as "June drop," but you may need to help the process along. For best production, apples should be about 6 inches apart, crab apples about 2 or 3 inches apart, and there should be a ratio of apples to leaves of about 30 to 40 leaves to one apple. Thin the fruit of dwarf trees to one apple to each spur; if this pruning stimulates too much leaf and stem growth, control the tree's size by root pruning.

MAPLE See *Acer*

MELIA (chinaberry, umbrella tree)

The chinaberry produces so many branches that although its leaves are small and fernlike, the tree will still provide dense shade. The branches are stout but brittle. Except when young, the chinaberry needs little training. Remove low branches while they are still small. To promote branching, shorten the ends of shoots to dormant buds that face in the direction new growth should take. Do maintenance pruning in the early spring, cutting off dead, winter-damaged or crossed branches. The Texas umbrella tree, although its branches grow upright at first, eventually assumes a low, dense, flattened crown without pruning.

MIMOSA See *Albizia*
MOCK ORANGE See *Philadelphus*
MONKEY PUZZLE See *Araucaria*

MORUS (mulberry)

Most mulberry trees are grown as shade trees, but the weeping varieties are grown as ornamentals and some of the smaller trees are used for hedges. All of them should be pruned in winter. And because most species grow rapidly—a 10-foot sapling will double its height in six years—trees

LEMOINE PURPLE CRAB APPLE
Malus purpurea lemoinei

First pruning   Second pruning   Third pruning

should be trained to the desired shape when very young.

As a shade tree, a mulberry can be trained to a single trunk or to several. For a single-trunk tree, select a central leader and shorten side branches at planting time. Mulberries intended for shade should be thinned out to open their dense top growth to air and light, and to provide headroom below the lower, often drooping branches. Weeping mulberries are seldom pruned except to remove crowded shoots. To train mulberries for a hedge, cut back the newly planted saplings every few weeks through the summer to encourage bushiness. As hedge plants, mulberries should simply be clipped to keep them shapely and within bounds.

MOUNTAIN ASH See *Sorbus*
MOUNTAIN LAUREL See *Kalmia*
MULBERRY See *Morus*

MYRICA (bayberry, wax myrtle)
The pungent green bayberry shrub bears its greenish flowers and aromatic gray berries on new wood. Species are either evergreen or deciduous, but even on deciduous versions leaves and berries linger late into the winter. Bayberry bushes are so compact and bushy that little pruning is required to keep them handsome and thriving once they are well established. But they respond well to hard pruning and can be successfully trained as small trees or clipped hedges.

Bayberries require only an annual thinning out of dead stems in the early spring. Suckers that appear at the base should be spaded out to curb the plant's spread. To train a tall plant as a tree, thin out all but one to four stems; then cut back or thin out side growth, and remove all suckers along the lower portions of the stems.

N

NANDINA (heavenly bamboo)
A slow-growing mild-climate shrub, heavenly bamboo produces graceful stems 3 to 8 feet tall that bear terminal clusters of white flowers in summer, followed by bright red berries in autumn. Foliage is deep green in summer and red or bronze in the fall, remaining bronze through early spring. Prune the plants in summer, cutting a few of the older stems to the ground each year. If a shrub becomes too tall, thin only the tallest stems. Straggly older shrubs can be rejuvenated by cutting all the stems back to a height of 6 to 12 inches; never cut them only halfway back, as this produces an unattractive plant.

NATAL PLUM See *Carissa*
NECTARINE See *Prunus*

NERIUM (oleander, rosebay)
Although these warm-climate flowering evergreens grow naturally as multiple-stemmed shrubs, they may be trained as informal hedges or as single-trunked tree standards in containers. They flower from late spring to fall on growth of the current season. Prune them in early spring to remove winter-damaged wood and to control their size and shape. Cut off spent flowers immediately to encourage more bloom. All parts of the oleander are poisonous, so dispose of cuttings with care; do not burn them, as even the smoke is toxic.

Oleanders grow vigorously, especially when pruned, and have a tendency to become leggy. To control height, cut back canes; to encourage bushiness pinch out the branch tips. Thin out the oldest canes every six or seven years. Rejuvenate badly shaped or overgrown plants by cutting them back to 6 inches above ground level.

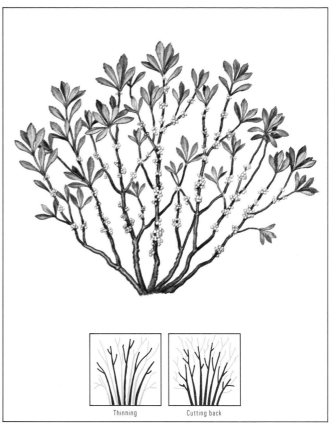

BAYBERRY
Myrica pensylvanica

OLEANDER
Nerium oleander

133

To train oleander as a standard, keep only one stem and cut off lower branches to the desired height. Pinch out the growing tips of the remaining branches to shape the top. Oleanders grown as standards tend to produce many suckers around the base; these too must be removed.

NEW JERSEY TEA See *Ceanothus*
NORFOLK ISLAND PINE See *Araucaria*

NYSSA (pepperidge, tupelo)

Of the two forms of tupelo cultivated in American gardens, the most common one is the pepperidge (the other tupelo is a swamp tree, grown mostly in the South). A tall-growing tree with strong wood, the sturdy pepperidge is conical when young but flat-topped at maturity; it is noted for its many zigzagging small branches which give the tree a picturesque silhouette in winter.

Young pepperidges need to be trained to a central leader by removing competing leaders and cutting back branches to half their length. For the first few years, to strengthen the tree's scaffold, remove some of the twiggy growth along branches. Once established, the pepperidge needs pruning only to remove crowded or dead branches.

O

OAK See *Quercus*
OLD-MAN'S-BEARD See *Chionanthus*

OLEA (olive)

This versatile evergreen may be grown as a tree, shrub or hedge and is ornamental as well as useful. Prized in Europe for its fruit, in the United States it is often pruned to show off its gnarled trunk and branches. All olive plants require regular corrective pruning, but those that are grown as shade trees are the easiest to maintain. Shrubs may bear more fruit and require less room but need more frequent pruning. If the fruit is not wanted (it stains patios and litters lawns), prune after the plant has flowered. Otherwise, prune after the harvest. To encourage more fruit, remove at least one third of the small branches annually, especially those that have recently fruited.

Olives grow naturally with a central leader. To train a tree form, retain and stake this leader and remove side branches below the point where you wish the branching to start. Remove suckers that sprout around the base. If the tree becomes too tall, cut back the leader.

Multiple-stemmed shrubs may be trained by removing the central leader on young plants, saving and staking several well-spaced branches or basal suckers. Olive hedges should be sheared each year to keep them dense. This regular shearing may prevent fruit production.

OLEANDER See *Nerium*
OLIVE See *Olea*
ORANGE See *Citrus*
OSIER, PURPLE See *Salix*

OSTRYA (hop hornbeam)

Slow-growing, with slender but strong spreading branches that are seldom injured by wind or ice, the hop hornbeam reaches a height of 40 to 50 feet, forming a deep irregular crown on a short trunk. If it is left to grow naturally, a tree may have either a single trunk or multiple trunks. It produces a heavy flow of sap in spring and should be pruned in winter in order to avoid bleeding. To train a young tree to a single trunk, cut back half the length of the side branches

First pruning   Second pruning   Third pruning

COMMON OLIVE
Olea europaea

134

during the first few years of growth and remove any vertical branches that threaten to dominate the leader. Thin out crossed and crowded branches.

OXYDENDRUM (sourwood, sorrel)

Developing naturally into a pyramid-shaped tree, in some cases with more than one trunk, the sourwood requires very little training or pruning when it is young and even less maintenance pruning in maturity. It is a small tree—its usual height is about 25 feet—and is very ornamental. It produces hanging clusters of fragrant flowers in midsummer followed by decorative seed pods in the fall. Prune sourwood in late winter or early spring.

To train young sourwoods, remove lower branches at the desired height, and thin out crossing or weak branches. On older trees remove dead, injured or weak branches, and any that have been damaged by extreme cold.

P

PARTHENOCISSUS (woodbine, American ivy, Boston ivy, Virginia creeper)

Using rootlike tendrils that grow along its stems, woodbine will quickly cover a fence or other garden structure with deciduous foliage. Thin it heavily at any time to remove deadwood and old, tough stems. Cut back loose, straggly shoots and train the vine to one layer. The holdfasts leave their remains on building woodwork or window screens, so cut the vine back before it reaches these surfaces. Once the vine has been pulled away from a support, it will not refasten itself; cut back the loose portion to a firmly holding stem. Do not attempt to rejuvenate old vines by cutting them back severely; woodbine does not respond to this treatment.

PEACH See *Prunus*
PEAR See *Pyrus*
PEARLBUSH See *Exochorda*
PECAN See *Carya*
PEEGEE HYDRANGEA See *Hydrangea*
PEPPERIDGE See *Nyssa*
PEPPER TREE See *Schinus*
PERSIMMON See *Diospyros*

PHILADELPHUS (mock orange)

This fragrant shrub, which blooms in late spring or early summer on wood formed the previous year, has naturally arching branches that usually are best left alone; if the mock orange is cut back, it tends to develop an ugly shape. The one exception is the Lemoine mock orange, whose less vigorous hybrids are stronger if canes are cut back to healthy shoots as soon as they have finished flowering, thereby forcing the plants to develop new growth to produce blossoms the following year.

To renew a mock orange and relieve crowding, remove some of the oldest canes at the ground every three or four years. Smaller, twiggy growth may also be thinned out to open crowded sections at the top.

PHOTINIA

The fast-growing photinias are generally treated as ornamental lawn shrubs, though taller species like the Chinese photinia are sometimes trained as small single-trunked trees. Some species are evergreen; others are deciduous. All have white flowers in spring, red berries in late fall, and handsome foliage. The foliage of evergreen photinias is bronze when young, while that of deciduous species turns scarlet in the fall. Prune photinias in late winter or early spring.

SORREL TREE
Oxydendrum arboreum

First pruning   Second pruning   Third pruning

FRASER PHOTINIA
Photinia fraseri

Thinning   Cutting back   Rejuvenating   Pinching

MOERHEIM COLORADO SPRUCE
Picea pungens moerheimii

JAPANESE ANDROMEDA
Pieris japonica

Allowed to grow naturally on multiple stems, photinias need little shaping. Thin to eliminate crowded branches and dead or damaged wood. Rejuvenate by cutting some of the older stems to the ground. For bushier growth, pinch off the tips of branches every time they add another 12 to 18 inches. To limit the size of the tall-growing species, cut back their top growth. To train photinia to a single trunk, remove all but one stem and all lower side growth.

PICEA (spruce)

Spruce trees, which may grow 1 to 2 feet a year when young and can become 75 feet or more tall, develop a naturally symmetrical shape with little more than occasional pruning of wayward branches. The rapid growth of some species is an asset in the creation of windbreaks and privacy screens, but for gardens that have limited space choose one of the slow-growing dwarf varieties, because spruces cannot be pruned to reduce their size.

Spruces need little pruning care. If more than one leader develops on a young tree, remove all but the strongest. A branch protruding beyond the tree's conical outline may be cut back to a bud in early spring. To slow the tree's growth and thicken branches, pinch off half of the soft, bright green new growth when it sprouts in spring. To maintain the soft, natural outline of a spruce, cut back extra-long branches far enough so the stubs will be concealed by other branches.

PIERIS (andromeda)

This broad-leaved evergreen shrub develops a naturally symmetrical shape and usually needs little pruning. Species range from 3 or 4 feet to as much as 10 feet in height and may be grown as single lawn specimens or in groups near borders. They are at their most attractive when allowed to grow in an uncrowded space.

Andromedas are many-stemmed plants that bloom in early spring from buds formed in the previous summer and fall. Prune them immediately after they flower. Pinch off spent blossoms to prevent seed pods from forming, and cut back branches that disrupt the plant's symmetry. As new shoots form from the base, pinch them back; this will encourage branching, although it may delay the development of the plant's natural shape and its flowers for the next season.

To rejuvenate mature shrubs, thin out a third of the oldest canes each year. You can also rejuvenate an overgrown or badly shaped andromeda by cutting the entire plant back to the ground. Do this in early spring to allow new growth to mature before autumn frost. In winter, unwanted branches may be pruned and forced into blossom indoors.

PINE See *Pinus*

PINUS (pine)

Pine trees are generally used as specimen trees, though low-growing dwarf species can be treated as shrubs. When young, many pines have broad pyramidal shapes, but they become rounder at the top and lose their lower branches as they age. In spring, a surge of new growth at the ends of the branches produces elongated new needle-covered shoots that are called candles.

Prune pines in the spring just as the candles begin to grow but before the new needles have fully developed. If young trees have more than one leader, select the strongest and cut back all others to their base. To make pines bushier and to slow their growth, either remove the central candle in each cluster or pinch off half of each candle. (If you remove the entire candle cluster, the branch will cease growing; repeated removal of all the candles on a branch will eventually kill the branch.) If a branch grows beyond the tree's symmetrical outline, cut it back to the point where side branches occur. But never cut pine branches back to wood containing no needles, since no buds exist there to form new shoots. Older pines can be made to appear taller and more treelike by removing some of the lower branches at the trunk.

PISTACHIO See *Pistacia*

PISTACIA (pistachio)

Broad-spreading pistachio trees have short, stocky trunks and heavy branches that tend to droop as the trees mature. They are grown as shade trees, as lawn specimens and on patios in tubs. The 30-foot common pistachio bears edible nuts in the fall on the previous year's growth; the taller Chinese pistachio bears ornamental fall berries and is sometimes used as grafting stock for common pistachios.

Prune pistachios in late winter before new growth begins. Stake each newly planted tree and select a leader and four or five well-spaced branches with wide crotches to serve as a scaffold, beginning about 4 feet above ground level. Older trees rarely require pruning except to remove damaged or broken branches and suckers. To keep tub-planted pistachios within bounds, prune back heavily in late winter.

PITTOSPORUM

These semitropical broad-leaved evergreens come in many forms—weeping, upright, mounded, spreading. They can be used in foundation plantings or as specimen shrubs or small trees. Left to grow naturally, pittosporums need only light pruning to maintain their shapes, but the shrub forms may be sheared into hedges, and the dense foliage of the Japanese pittosporum may be thinned and clipped into tiers and other topiary patterns. On all forms, when pruning, cut back to just above a node. Pittosporums can be pruned at any season; when they are grown as hedges they may need to be sheared as often as three times a year—in spring, in midsummer and again in early fall.

Prune all forms of pittosporum to shorten long branches and to preserve the shapes of the plants. Remove sucker-like growth that springs from the branches and trunks of some species, spoiling their lines. Cut away dead and damaged branches. On tree forms, remove lower branches for clearance. Remove the occasional all-green shoots that mar the effect of variegated pittosporums; if allowed to remain, these shoots will eventually dominate the plants.

PLANE TREE See *Platanus*

PLATANUS (plane tree, sycamore)

The plane tree, valued for its mottled bark and its ability to survive in urban gardens, grows rapidly and eventually reaches a height of 100 feet or more. Mature trees require little pruning, except to remove dead, diseased or damaged branches. A young tree should be pruned to encourage a strong, straight trunk. Cut back newly planted young trees to half their height. Remove side shoots at the end of each summer until the tree is 6 feet tall. Then, as the tree matures, thin occasionally to reveal the plant's interesting bark and beautiful form.

The London plane tree can be pollarded or pleached to form a low, dense canopy of foliage. To pollard, cut off scaffold branches about two feet out from the trunk after the leaves have fallen. Each callused stub will produce many new branches the following spring; cut these back again each

EASTERN WHITE PINE
Pinus strobus

Pinching candles   Cutting back

JAPANESE PITTOSPORUM
Pittosporum tobira

Cutting back   Shearing

LONDON PLANE TREE
Platanus acerifolia

First pruning   Second pruning   Third pruning

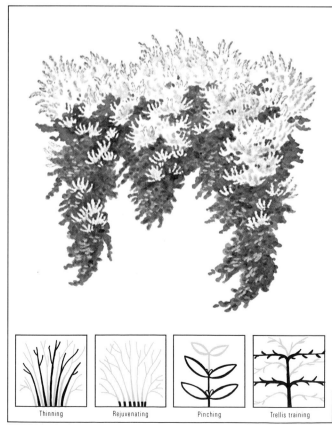

SILVER LACE VINE
Polygonum aubertii

Thinning   Rejuvenating   Pinching   Trellis training

fall. To pleach, twine branches of at least two trees onto an arched framework of wood or metal, cutting off extraneous growth that extends beyond the frame.

PLUM See *Prunus*

PODOCARPUS (fern podocarpus, yew podocarpus)

This warm-climate narrow-leaved evergreen grows from 6 to 60 feet tall, depending on the species, but all species have a similar pattern of growth. Podocarpuses are narrow trees with horizontal branches that in maturity droop gracefully at the tips. Unlike most evergreens, they grow continuously all summer. They may be used as individual lawn specimens or in formal or informal hedges, or they may be trained and pruned into elaborate topiary and espalier designs.

Cut away broken or dead branches in early spring. You can prune podocarpus to shape and train it at any time, though early spring to midsummer is best. To train young trees, select one to four strong stems and remove the weaker ones. Pinch back new growth to encourage bushiness. As a plant matures, prune to produce the desired shape. For compact plants, prune back half of new growth. On tall-growing trees, thin out some branches to expose the form and bark of the trunk. For informal hedges, trim out unwanted branches by reaching into the interior of the plant and cutting where the branch joins another branch. For formal hedges and topiary work, shear as desired.

POINSETTIA See *Euphorbia*

POLYGONUM (silver lace vine)

Of the 150-odd species of polygonum, the one most commonly cultivated is the silver lace vine, a handsome, sturdy deciduous plant that grows exceptionally fast (up to 30 feet annually), producing masses of fluffy white flowers all summer long on new growth.

Prune silver lace vine in late winter or early spring, while the plant is still dormant. Remove winter-damaged wood. Cut back main and side stems to the third or fourth bud of the previous year's growth; then thin out side shoots. In very cold climates, where silver lace vine freezes down to the ground, cut back the deadwood to ground level; new growth will sprout from the roots, but blossoming may be delayed until late summer. During the summer, the silver lace vine's rampant runners can be kept under control by pinching back or lightly cutting back new growth.

POMEGRANATE See *Punica*
POPLAR See *Populus*

POPULUS (poplar)

Poplars are fast-growing short-lived trees that will tolerate heavy top pruning or root pruning to keep them in bounds. But both procedures stimulate production of root suckers, which should be removed immediately. All species of poplar are brittle, and a broken limb should be pruned to a crotch. Prune poplars in summer or fall to avoid bleeding.

Columnar Lombardy poplars make handsome specimens or, planted close together, tall screens or windbreaks. To stimulate their characteristic low branching, cut back newly planted saplings to one foot, then reduce new growth by half for the first year or two.

The spreading white and eastern poplars, often grown as shade trees, should be shorn of their low-growing branches when young to provide headroom. Both should be thinned about every 10 years to improve their somewhat untidy

138

habit of growth, but on the white poplar this thinning should be confined as much as possible to small branches, because cuts leave black scars on its silvery bark.

POWDER PUFF See *Calliandra*
PRINCESS FLOWER See *Tibouchina*
PRIVET See *Ligustrum*

PRUNUS (apricot, cherry, nectarine, peach, plum)

Most species of these stone-fruit trees are selected for either fruiting or flowering characteristics and pruned accordingly. As ornamental flowering trees, they are trained when young to single leaders and thereafter are pruned only to keep them looking attractive. Thin out dead or crowded branches in winter while trees are dormant; cut back side branches to shape trees in the late spring after flowering.

For fruit production, stone-fruit trees are usually planted in the fall in warm climates, in spring where winters are severe. In both cases, they are pruned immediately. Remove all but three to five well-spaced branches that form angles of more than 45° with the trunk. Cut these branches back to about 10 inches. The following summer, remove all but two end shoots from each of these branches. The third year, remove half the previous season's growth. Thereafter, pruning varies with the species and how it bears fruit.

Apricots are borne on stems from growth of the previous year and on spurs two to four years old. Prune the tree in late winter, cutting back the older branches to younger side branches one third to halfway in from their tips. When fruits appear, thin them to 3 inches apart to prevent heavy crops from breaking the branches.

Cherry trees produce their crops on fruiting spurs up to 12 years old. In their third and fourth seasons, cut upright shoots back to lateral shoots, since vertical shoots produce no fruit. Thereafter remove about 10 per cent of the oldest wood each winter, stimulating the tree to replace it with more vigorous growth.

Peaches and nectarines bear fruit on the previous year's growth. Cut back all year-old growth by one third in winter and thin fruit in early summer to about 10 inches apart.

Plums bear fruit on six- to eight-year-old spurs. Cut back branches one quarter of their length each winter and thin immature fruit to about 6 inches apart.

Fruit trees should be pruned each winter to remove dead, crossed or crowded branches; suckers that develop at the bases of the trees should be removed each spring.

PSEUDOTSUGA (Douglas fir)

These tall-growing conical evergreens are planted singly as accent trees or in clumps as windbreaks and hedges. When young, they can be pruned to make them bushy, but this training ceases after they are about 6 feet high. Mature trees are pruned to remove dead or damaged wood and those grown as hedges are generally sheared annually. Shearing should be done early in the spring, so that new growth will quickly hide the clipped stubs.

To train a young seedling, cut back its branches until a strong central leader develops; if two leaders form, retain the most promising one and remove the other. When the tree is about 18 inches high and until it is about 5 feet high, pinch off half the new growth at the tips of its branches every year to make the tree bushier.

PUNICA (pomegranate)

The bushy, deciduous pomegranate is a shrub that can be trained as a small tree. It is an ornamental plant used singly,

First pruning   Second pruning   Third pruning

KWANZAN CHERRY
Prunus serrulata 'Kwanzan'

Thinning   Cutting back   Shearing

POMEGRANATE
Punica granatum

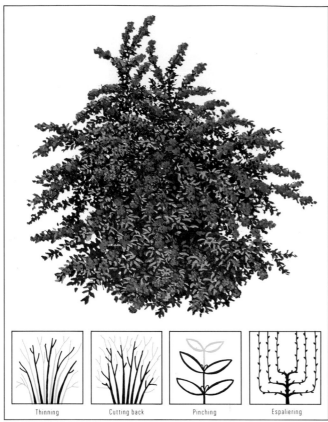

LALAND FIRE THORN
Pyracantha coccinea lalandei

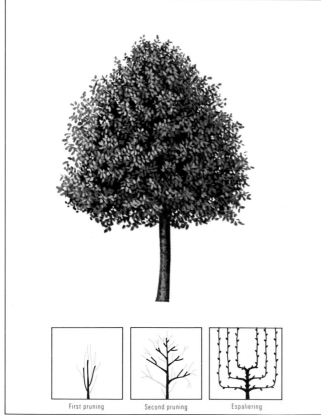

BRADFORD PEAR
Pyrus calleryana 'Bradford'

or it can be used in a hedge. Single-flowered pomegranates bloom in the summer, then produce crops of scarlet-fleshed, slightly acid fruit in late summer or fall. Non-fruiting pomegranate varieties produce enormous flowers up to 4 inches across with double rows of petals that last for several months in the summer.

Since pomegranate flowers and fruits are borne on the current season's growth, these plants should be pruned in early spring before the buds open. For a tree, train a young plant to a single leader and remove all but several well-spaced, wide-angled side branches. As the tree develops cut back branches slightly to promote vigorous new growth. Thin damaged or weak wood and remove suckers at the base of the plant each spring.

For pomegranates that are grown as shrubs, remove about one quarter of the old stems at ground level every spring. Cut new growth back slightly to stimulate profuse blooming. Clip off shoots that extend beyond the plant's natural shape. Hedge specimens can be sheared, but you should take care to remove as little of the new growth as possible so that the flower crop is not destroyed. Remove interior twigs when a shrub becomes too thick.

PURPLE OSIER See *Salix*

PYRACANTHA (fire thorn)

A fast-growing broad-leaved evergreen shrub that bears tiny white flowers in late spring and clusters of bright red or orange berries in the fall, the fire thorn will withstand heavy pruning. It can be espaliered against a wall or fence, trained as a tree standard or clipped into a hedge or topiary design. Branches laden with berries can be cut in early winter for indoor decorations.

Control the shape of young fire thorns by pinching and cutting back new growth in spring and summer. To produce compact clusters of fall and winter berries, cut back new growth after the plant blooms to just above the first flower cluster. Cut old leggy stems to the ground in late winter. Do not leave any stubs, as they may invite rot.

PYRUS (pear)

Pear trees are extremely long-lived, bearing fruit for as long as 100 years. Two, the evergreen pear and Callery pear, are grown exclusively as ornamentals because their fruit is inedible, but many of the edible-fruited varieties are also grown as lawn trees.

The evergreen pear, if left unpruned, becomes a many-stemmed shrub with long drooping branches; it is usually trained to a single trunk with shortened branches, though it may also be espaliered. The Callery pear seldom needs more than maintenance pruning except to remove lower branches to provide more headroom.

Pruning for edible-fruited pears is based largely on improving the quality and ensuring the accessibility of the crop. Standard-sized pear trees can be kept pruned to a height of about 20 feet, so their fruit is accessible from a stepladder. Dwarf pears, which are only 8 to 10 feet high at maturity, can be espaliered.

In spring, at the time of planting, cut the young pear tree back to 3 feet and shorten its branches by half. Train the branches to emerge from the trunk at wide angles to it and as the tree grows, thin it so branches are spaced spirally around the trunk, about 6 to 12 inches from each other. Two or three years after planting, cut back the leader of a standard-sized tree to encourage side growth and to keep the fruit within reach. Cut back dwarf pears to keep the trees well-shaped and symmetrical. Any additional pruning should wait until the trees bear fruit—at four or five years for standard-sized trees, two or three years for dwarf trees.

When pear trees reach bearing age, thin the fruit in early summer after the trees have completed their normal fruit drop; save the most promising pear in each cluster and space the pears 6 to 8 inches apart. Prune the trees in winter or spring during their dormant period. Thin out the tops of the crowns to keep them open to light and air. Cut back some of the upright limbs to outward-facing buds, so that the fruit remains accessible. Remove any shoots that threaten to compete with the central leaders.

Q

QUERCUS (oak)

This huge group of shade and specimen trees includes the evergreen live oaks and the sturdy deciduous oaks, noted for their tough wood. Except for training when young and maintenance pruning, most oaks need little care. In fact, the branches of a mature oak should not be cut back because ugly sprouts tend to rise from the stubs.

To train a young oak, cut back its branches by a third until a strong leader develops. If the leader forks, which sometimes happens on such species as the white oak, cut off one of the branches of the fork. Prune a mature oak to remove dead or damaged branches, cutting back to a fork or to the trunk of the tree. Some oaks that branch very near the ground may have to be pruned to provide headroom if they are grown as shade trees.

Both the evergreen holly oak and the deciduous shingle oak are sometimes planted as hedges and windbreaks and may be sheared. The shearing should be done preferably before new growth starts in the spring, but light shearing may also be done during the growing season.

R

RAPHIOLEPIS (yedda hawthorn)

This glossy broad-leaved evergreen shrub, handsome in mild-climate gardens, is characterized by its slow growth. When mature, it is 4 to 6 feet tall and about 6 feet in diameter. It has multiple branching stems and bears fragrant white flowers in spring on growth of the previous year, followed by blue-black berries in summer.

Prune the yedda hawthorn soon after flowers fade. For bushier plants, pinch back the tips of branches. Pinch out low side shoots to encourage the shrubs to grow upwards.

RASPBERRY See *Rubus*
RED BAUHINIA See *Bauhinia*
REDBUD See *Cercis*

RHODODENDRON

With their annual display of funnel-shaped flowers clustered above rosettes of glossy, oval leaves, rhododendrons make striking informal hedges or specimen shrubs. Some species and hybrids of these broad-leaved evergreens have an erect habit of growth, while others are more spreading and may have a loose, open structure. All grow slowly to heights of 15 feet or more.

Rhododendrons form buds in summer and early fall that flower the following spring. Remove faded flowers, taking care not to damage the foliage buds just below them. These buds, when they sprout, determine the size and shape of the plant. To control young plants, pinch or cut back shoots from the foliage buds to two or three leaves or remove foliage buds entirely. If a branch must be cut to help shape a plant,

RED OAK
Quercus borealis

EVERESTIANUM CATAWBA RHODODENDRON
Rhododendron catawbiense 'Everestianum'

SHINING SUMAC
Rhus copallina

BLACK LOCUST
Robinia pseudoacacia

make the cut just above a leaf rosette or just above one of the faint growth rings that mark the location of a leaf from a previous season; a new shoot should then grow from this old growth ring. If the cut removes more than a few years' growth, a new shoot may not form.

Mature rhododendrons need little maintenance pruning except to remove dead or diseased wood, but older, straggly plants may need rejuvenating. Do this over a period of several years, fertilizing the plants heavily the year before starting to prune. In later winter of the first year, thin out one third of the old growth, choosing the oldest, weakest stems and cutting them back to 6 to 12 inches above the ground. Remove another third of the old growth in late winter of the second year, and complete the process in the winter of the third year. The new growth that sprouts from these stubs will become the future plant.

RHUS (sumac)

Sumacs, prized for their autumn foliage and bright seed heads, spread rapidly from suckers that are produced by underground stems. Low-growing species are often used as ground covers, others become massive shrubs, and the tall staghorn sumac grows to resemble a small tree. Sumac is generally pruned in early spring while it is dormant; its fruits and its inconspicuous spring flowers appear on new wood.

Sumacs need to be root-pruned annually to keep them from sending up suckers and spreading out of bounds. Cut back the branches of low-growing sumac when they become too long; cut back the branches of shrub sumac to keep the plant shapely. Remove crowded branches and remove a few of the oldest stems each year to encourage new growth. To enhance a staghorn sumac's treelike shape, remove all but the strongest stem and all low-growing branches.

RIBES (currant, gooseberry)

Thorny currants and gooseberries are planted as ornamental shrubs as well as for fruit. Both plants have attractive flowers and grow 2 to 6 feet high, spreading equally wide. Some species are evergreen, but most are deciduous.

Ornamental currants and gooseberries should be cut back to 6 or 8 inches from the ground when they are planted in the fall. Since the tubular red or yellow flowers appear on old growth, prune decorative shrubs in spring after flowering so the plants have time to develop buds for the next season's bloom. Cut back straggly side shoots and thin out weak or crowded stems. Currants can be sheared as hedges, but this will interfere with flower and fruit production.

For fruit production, young currant or gooseberry bushes should be thinned at the time they are planted to all but six well-spaced, strong canes. The following winter, remove all but the six original canes plus the three strongest new ones. Clusters of quarter-inch white, pink or red fruits will appear on old growth in the summer of the second year. Thereafter, thin the plants each winter to about 12 canes, saving four from each of the previous three years' growth and removing all canes older than three years.

ROBINIA (locust)

Distinguished by feathery leaves and pendulous clusters of early summer flowers, locust trees develop—with some pruning help from the gardener—into tall, slender specimens, useful more for ornament than for shade. The white blossoms are borne on the previous season's growth and are followed by smooth, flat seed pods that linger on the rather brittle branches most of the winter. Locusts bleed heavily in spring, so pruning should be restricted to late summer or fall.

To encourage a strong trunk, remove the slender whiplike branches along the trunks of young trees each year; these tend to grow horizontally at first but gradually follow ascending lines as the trees mature. Remove suckers as they appear, but avoid any disturbance of the roots, as this encourages even more suckers. As locust trees mature, some main limbs may become curiously contorted, as if they are straining outward; if they do not crowd or cross other branches, they may be retained for their peculiarly interesting growth patterns. Old locusts need little pruning except for the removal of dead or diseased limbs and any branches that have been damaged by wind or winter weather.

ROSA (rose)

Climbing roses, including ramblers, and bush roses are pruned differently once they are established. But all are pruned heavily when they are planted if they are planted bare-rooted. Cut back the strongest shoots of the new plant to 6 to 10 inches to balance the root system, and remove the weaker shoots. Make each cut at an outward-facing bud, at an angle of 45 degrees to the axis of the stem.

Familiar bush roses like hybrid teas, floribundas, hybrid perpetuals and grandifloras should be pruned in early spring before growth begins. Cut back all winterkilled canes to live wood, about ¼ inch above an outward-facing bud. Make sure that the wood is healthy by examining the cut for any discoloration. Remove any crossed branches. Beyond this point, you can prune very little (up to half the previous year's growth) or moderately (up to half the total length of all canes). The former method will produce a large bush with blossoms of moderate size. The latter will produce fewer but larger roses. Rejuvenate old bush roses by removing one quarter of the oldest canes at ground level every year for a four-year period.

Fence- or wall-trained climbing roses bloom on canes of the previous seasons and should be pruned soon after flowering has ceased. Thin out some of the old canes that have bloomed. As new canes develop in their places, cut them back during the growing season to a size and shape appropriate to the supporting structure.

To encourage more flowers on the large-flowered ever-blooming varieties of climbing roses, cut back each branch after it has flowered to ¼ inch above the second leaf cluster from the base of the branch. On the large-flowered, once-blooming climbers remove the spent flower clusters but cut back each branch only to the nearest leaf, leaving the rest of the branch for next year's flowers. Rejuvenate climbing roses by thinning out two of the oldest canes each year.

A tree rose, which is a bush rose grafted atop a bare trunk, which is in turn grafted onto a sturdy understock, requires pruning during its first few years to remove growths that may develop along the trunk and from the roots. Thereafter, prune to remove deadwood and diseased branches from the top every spring, and cut back the canes to train the top into a symmetrical ball.

ROSE See *Rosa*
ROSE APPLE See *Eugenia*
ROSE OF SHARON See *Hibiscus*
ROSEBAY See *Nerium*
ROSEMARY See *Rosmarinus*

ROSMARINUS (rosemary)

In warm climates, rosemary is a twiggy evergreen garden shrub that grows to a height of 3 feet or more and is sometimes shaped into small topiary figures. In the north, it is

Thinning Cutting back Disbudding Removing spent flowers

HYBRID TEA ROSE
Rosa

Thinning Cutting back Disbudding Removing spent flowers

CLIMBING ROSE
Rosa

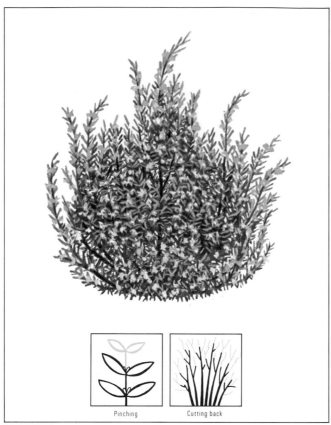

ROSEMARY
Rosmarinus officinalis

BLACK SATIN BLACKBERRY
Rubus hybrid

raised as a kitchen herb, its fragrant leaves used fresh or dried in cooking.

Plants grown for cookery or topiary purposes are trimmed as required. Otherwise, rosemary should be pinched when young to control its shape and cut back lightly in maturity to encourage new growth. Plants that are overgrown or are injured by cold are cut back in spring.

RUBUS (black raspberry, blackberry, raspberry)

Heavy crops of succulent summer berries reward gardeners who prune their raspberry, black raspberry and blackberry bushes. Pruning also keeps these bramble shrubs from becoming unsightly tangled thickets. Except for the everbearing varieties, the shrubs produce fruit on two-year-old canes, which then die.

Berries are pruned according to the manner in which they are grown, in rows or in clumps called hills. But the initial training for new plants is identical. Cut back the canes to 4-inch stubs and allow the new canes that sprout to grow undisturbed through their first summer. The second spring, begin the annual thinning out and cutting back that are needed for good fruit production, removing all the canes that have borne fruit.

Yellow or red raspberries grown in rows should be thinned to about three canes per foot, those in hills to six or eight canes. Canes that remain should be cut back to 3 or 3½ feet. With everbearers, cut the canes to the ground in the fall and allow unrestricted growth the following summer for a heavy crop the following fall.

Black raspberries and blackberries are also thinned to three stems per foot in rows, or six to eight stems per hill, and the canes are cut back. But the stems of these plants develop side shoots, which should be cut back to 12 inches at the same time the stems themselves are shortened.

RUSSIAN OLIVE See *Elaeagnus*

S

SALIX (willow, purple osier)

The rapid-growing, brittle-branched willows may be divided into two groups: trees such as the weeping willow and the more erect black willow, and twiggy, clump-forming shrubs such as pussy willows and the purple osier.

Prune tree willows in the summer or fall; they may bleed excessively if pruned in winter or spring. Remove dead branches and water sprouts that grow along branches as well as suckers and sprouts around the bases of the trees. To control the size of a tree, remove entire branches; willow trees, if cut back partway, may produce unsightly sprays of new growth from the cut end. To train a young willow to a single trunk, remove all but one stem and a few well-spaced side branches—enough to encourage the development of the trunk. In training weeping willows, remove lower branches that would trail on the ground.

Prune shrub willows in late spring, immediately after they have flowered, to allow time for the development of buds for the next year's catkins, which are produced on year-old wood. Shrub willows grow rampantly, and may need thinning at the center when branches become crowded. To rejuvenate an old or overgrown shrub willow, cut the entire plant back to the ground; in two or three years it will renew itself.

SAMBUCUS (elderberry)

These fast-growing deciduous shrubs and trees, which bear clusters of tiny creamy-white flowers in early summer and edible blue-black or red berries in early fall, can be pruned as heavily as desired without being injured. Prune in winter or early spring before new growth starts so that the current season's growth can produce flowers and fruit.

Blue elderberry can be grown as a shrub or trained as a multitrunked tree. To grow it as a shrub, cut the plant to the ground each winter, forcing new canes to appear the following spring. To grow it as a rounded tree with slightly weeping branches, remove all but one to four canes and any suckers that subsequently develop. Thin out surplus twiggy branches in late winter.

Young canes of American elderberry should be pinched back in winter to force additional branches to form and subsequently bear fruit. In fall, thin out two- and three-year-old canes that are no longer productive.

SASSAFRAS

This slender upright aromatic tree is decorative by nature and needs little pruning to maintain its shape. It is generally grown singly as a specimen plant but may be deliberately stunted for decorative use. Its rate of growth may be rapid when young, between 2 and 3 feet a year, and its eventual height is 20 to 60 feet. The trunk is heavy and the branches short and horizontal. In spring, the branches bear clusters of sweet-smelling yellow flowers; in fall, the odd foliage, of three different shapes, turns scarlet, gold and orange.

Sassafras branches are brittle and may suffer winter damage. Prune the tree in late winter to remove dead or broken limbs. In spring, remove any suckers that grow around the base of the tree. These can generally be pulled off the roots easily; otherwise, cut them away, but disturb the roots as little as possible, for this encourages more suckers.

Left to grow untrained, a young sassafras will develop an open, asymmetrical crown. In dwarfing, this look is deliberately heightened. You can cut back new growth severely to limit the tree's size; the selective removal of branches may be used to create a windswept silhouette.

SCHINUS (pepper tree)

These ornamental evergreen or semideciduous trees are grown in warm, dry climates as shade trees and for their clusters of bright wintertime berries. The two most commonly used species, the California pepper and Brazilian pepper, are different in appearance but have similar pruning needs. Both are roundheaded and spreading and may have multiple stems, but the California pepper tree grows 20 to 50 feet and has gracefully drooping branches and a short gnarled trunk, while the Brazilian pepper tree is more upright and compact and grows only 25 feet high; it is often grown in a pot on a patio. Both have brittle branches subject to storm damage.

Prune pepper trees in late winter or early spring. Cut back long limbs and remove some limbs to minimize storm damage. Remove low branches to provide headroom. Pepper trees are usually permitted to develop several trunks, but the California pepper is sometimes trained to a single trunk by removing all the stems except one and cutting back lateral branches for several years until the trunk is well established.

SERVICEBERRY See *Amelanchier*
SHAD-BLOW See *Amelanchier*
SHARP-LEAVED JACARANDA See *Jacaranda*
SHE-OAK See *Casuarina*
SHRUB ALTHEA See *Hibiscus*
SILK TREE See *Albizia*
SILVER LACE VINE See *Polygonum*
SNOWBELL See *Styrax*
SNOWBERRY See *Symphoricarpos*

BADYLON WEEPING WILLOW
Salix babylonica

First pruning　　Second pruning　　Third pruning

CALIFORNIA PEPPER TREE
Schinus molle

First pruning　　Second pruning　　Third pruning

JAPANESE PAGODA TREE
Sophora japonica

First pruning   Second pruning   Third pruning

EUROPEAN MOUNTAIN ASH
Sorbus aucuparia

First pruning   Second pruning   Third pruning

SOPHORA (Japanese pagoda tree)

Large, spreading and roundheaded by nature, this lacy-leaved shade tree needs little pruning to keep it in shape. This is particularly true of the weeping pagoda tree, whose pendulous contorted branches give it a special charm. Prune pagoda trees to remove dead or damaged branches and, if a tree is being used for shade, remove any branches that threaten to interfere with headroom. In maturity the tree's multiple stems may become crowded and require thinning. The same is true of its crown, which may become too dense to let light and air into the interior.

To train a young tree to a single trunk, remove all but one stem and remove or cut back side branches for a few years until the trunk is established and is about 3 inches in diameter. Then allow the tree to develop naturally.

SORBUS (mountain ash, whitebeam)

The most familiar of the bright-berried sorbus trees is the ornamental mountain ash, but the genus also includes such large shade trees as the wide-crowned whitebeam. All grow rapidly to heights from 20 to 60 feet and bear great terminal clusters of flowers in early summer and berries in early fall. The trees should be pruned in later winter or early spring.

Sorbus trees are usually allowed to develop naturally with one or more trunks. Prune them to remove dead or damaged branches and branches that cross or crowd each other. When used as a shade tree, sorbus is generally trained to a single trunk. Until the tree has reached a height of about 15 feet, thin out any competing leaders and lightly cut back all lateral branches once a year. Remove any branches that form narrow crotches at the trunk, and any low-growing branches that interfere with headroom.

SORREL See *Oxydendrum*
SOURWOOD See *Oxydendrum*
SPICE BUSH See *Lindera*
SPINDLE TREE See *Euonymus*

SPIRAEA (bridal wreath)

Spireas are pruned according to their blooming time and their habit of growth. One group of these popular shrubs blooms in the spring on wood of the previous season and should be pruned right after the plants have flowered. Another group blooms in the summer on new wood and should be pruned in early spring. The spring-blooming plants produce mostly white flowers, while those that bloom in summer may be white, pink or red. Both groups come in many forms: tall and upright, slender and arching, low-growing, dense and almost globular.

Prune upright spireas to remove deadwood and old, nonproductive stems. Cut out dead twigs and cut back branches to keep a plant shapely, encouraging new shoots and bushier growth, especially at the base. Prune arching spireas at ground level when canes become old or overcrowded, but prune top growth as little as possible, as this tends to spoil the plants' shapes. Prune mound-type spireas by pinching them back to keep them compact. Snip out deadwood and twiggy overgrown shoots. All spireas produce many suckers, which should be removed unless you wish to replace old stems with new growth.

SPRUCE See *Picea*

STEWARTIA

The most commonly grown forms of the summer-flowering stewartia are large shrubs, but a few of them are small trees.

The tree types rarely surpass 30 or 40 feet in height in gardens and usually assume a rounded, low-branching form. Shrubby forms of stewartia grow from 10 to 15 feet tall. Both types are open branched, revealing the characteristically flaky bark on their trunks.

These moderately slow-growing plants develop into shapely specimens with little more than maintenance pruning in late winter or early spring. Cut back overlong shoots that mar the plants' natural shape, and thin out crowded inner branches. Remove diseased or damaged branches by cutting back to buds or branches on healthy wood.

STRAWBERRY BUSH See *Euonymus*
STRAWBERRY TREE See *Arbutus*

STYRAX (snowbell)

The snowbell is a wide-spreading shrub or small tree that grows naturally on multiple stems but may be trained to a single trunk. In early summer after the leaves have developed, it bears clusters of pendulous white, sometimes intensely fragrant flowers on the previous season's growth. Prune it in late winter or early spring; some flowers will be lost, but summer pruning would promote growth unable to harden before frost.

Snowbells, when they are mature, have a rather open, flat-topped look. Tree forms generally reach from 20 to 30 feet in height and achieve an equal spread; shrubs usually become 6 to 9 feet tall. Either form may be trained to a single trunk, or leader, by selecting a strong stem and removing all competing stems. As lateral shoots develop, thin out those that threaten to crowd each other. Snowbells may need pruning to control their lateral spread and preserve their shape. Cut back long side branches as necessary, and thin out crowded branches to keep the plants open.

SUMAC See *Rhus*
SURINAM CHERRY See *Eugenia*
SWAMP CYPRESS See *Taxodium*
SWEET BAY See *Laurus*
SWEET GUM See *Liquidambar*
SWEET SHRUB See *Calycanthus*
SYCAMORE See *Platanus*

SYMPHORICARPOS (snowberry)

These ornamental shrubs, grown primarily for their clusters of waxy fall berries, are generally loose and spreading, with multiple stems and upright branches that arch gracefully to the ground when laden with fruit. The berries develop from small flowers that bloom in early summer on the current season's growth. Plants should be pruned in early spring while still dormant.

Despite their haphazard shape, snowberry bushes should not be sheared; cutting back branches partway tends to produce an abnormal twiggy growth. Every year, dead, damaged or crossing branches should be removed entirely, back to the nearest healthy crotch or to the ground; unwanted suckers should also be removed. Every three years or so, thin out the old and weak canes at ground level and trim out any branches that mar the plant's shape, which should be well groomed but informal.

SYRINGA (lilac)

The lilacs, a large group of shrubs that are grown chiefly for their profuse clusters of often highly fragrant spring flowers, bloom on the wood of the previous year. Many are hybrids, and they come in a wide range of sizes and shapes,

BRIDAL WREATH
Spiraea prunifolia

VICTOR LEMOINE LILAC
Syringa vulgaris 'Victor Lemoine'

from tall and treelike to small and rounded, from arching to upright, and from dense to spreading. Large lilacs are grown as single specimens and in groups as screens and windbreaks; smaller ones serve as informal hedges.

Lilacs grow vigorously as they mature but may start slowly; do not prune young plants for four to five years after planting except to remove broken shoots—and on grafted plants, any shoots that appear below the knobby point on the grafted stem. After flowering has started, cut off spent flower heads—back to the first leaves below the clusters—before they have a chance to form seed heads. On mature lilac plants, thin out several of the oldest stems each year, as well as any weak stems. Remove new shoots that develop at the base of the plant unless a few are needed to replace old stems, in which case select three or four strong, well-spaced ones and remove the others.

To relieve overcrowding, remove enough stems and basal shoots to open up the interior of each plant. To rejuvenate an old overgrown lilac, cut back old stems to the ground, and remove all but six or so of the most vigorous younger stems; the following spring, these may be cut back to a height of 5 or 6 feet. Alternatively, rejuvenate by cutting back the entire plant to 4 to 10 inches above the ground; it should recover and bloom again in three or four years.

T

TAMARACK See *Larix*
TAMARISK See *Tamarix*

TAMARIX (tamarisk)

The arching, feathery tamarisk shrubs grow on multiple stems to a height of 8 to 15 feet, and strong species like the French tamarisk can be trained to small, single-trunked trees. Spring-flowering plants, such as the small-flowered tamarisk, bloom on growth formed in the previous season; summer-blooming species, such as the five-stamened tamarisk, flower on new growth.

Spring-flowering tamarisk should be pruned immediately after the flowers fade. Cut back the branches that have flowered, encouraging new shoots to grow in their places. Prune summer-flowering species in early spring while the plants are still dormant. Cut back their stems to two or three buds and remove a few stems entirely. This severe pruning should result in a multitude of healthy new shoots, from which a clump of four to eight stems should be permitted to develop. Because fast-growing tamarisks tend to sprawl, they may be radically cut back to the ground if this becomes necessary to keep their growth under control.

To train tamarisk as a small tree, remove all but one stem. Cut out all suckers that develop at the base of the plant and thin side shoots on the stem to establish a scaffold of several strong branches with wide crotches.

TAXODIUM (bald cypress, swamp cypress)

Although called a cypress, this tree is actually a member of the pine family, and to compound the confusion, it is deciduous. Its soft, feathery needles drop off in the fall and the tree remains bare, except for its tassel-like male flowers and tiny globular cones, until the following spring. In its native swampy habitat, the bald cypress becomes very large, up to 150 feet high with a spread of 20 to 25 feet, and is prized as a timber tree; the wood is very durable. In northern gardens it is a smaller tree, 50 to 70 feet high, and is grown for its decorative value. It develops a tall, straight, tapering trunk and a narrow pyramidal shape that becomes more spreading with age. It lives for hundreds of years.

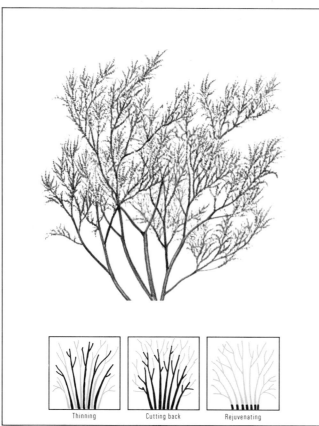

FIVE-STAMENED TAMARISK
Tamarix pentandra

Prune the bald cypress in the spring. If newly planted young trees develop more than one leader, select the strongest and cut off any competing leaders. To encourage fuller, bushier trees, cut back the tips of the branches for the first three or four years. Mature trees need no pruning other than removing deadwood or damaged branches.

TAXUS (yew)

Because slow-growing yews tolerate shearing extremely well, they are favorite plants for formal hedges and topiary. From low creeping ground covers and spreading shrubs to globe-shaped, columnar or conical trees, these evergreens can provide an enormous selection of plant material for landscaping. Dwarf and semidwarf varieties of the Japanese and English yews are commonly available and can be kept to a given size and shape almost indefinitely with annual pruning.

Young plants need little more than light shearing or pinching in early spring before new growth begins. If the foliage becomes too dense, branches that fall beyond the desired outline may be cut back to a fork. Shear hedges and topiaries in early spring, again in midsummer, and a third time in midfall if necessary to maintain their lines.

To rejuvenate old yews, thin out crowded branches and cut back remaining branches severely in early spring. New shoots developing from dormant buds on old wood will quickly hide branch stubs.

TEA TREE See *Leptospermum*

THUJA (arborvitae, northern white cedar, western red cedar)

With their flat sprays of scalelike leaves, durable arborvitae trees and shrubs have become a staple in home landscaping. New growth is characteristically vertical, but only Oriental varieties maintain this silhouette; the sprays on older branches of other arborvitaes become horizontal. Not all are completely evergreen; varieties of *T. occidentalis* turn brown in the winter in colder areas. Giant arborvitae, also called western red cedar, and northern white cedar, an arborvitae despite its common name, grow naturally into towering conical or pyramidal trees. However, numerous dwarf and semidwarf varieties of various species have been developed for landscaping use. Globular, conical or pyramidal, all of these smaller, shrub-type plants tolerate considerable shearing.

Arborvitae trees need little or no pruning unless competing leaders develop. In this case, remove the weakest one. The growth of giant arborvitae can be slowed by root-pruning them, severing the roots about one third of the way in toward the plant's base in early spring before growth begins.

For shrubs, several vertical leaders are desirable for bushiness. Allow the multiple leaders of young arborvitaes to develop until the shrubs reach the desired height. Shear or cut back any ragged side branches in early spring before new growth begins. Once the shrubs reach the desired size and shape, maintain them by shearing off both the new top and side growth in late spring or early summer.

To rejuvenate overgrown arborvitae shrubs, cut them back or shear them severely in early spring so that the new growth will hide the branch stubs.

TIBOUCHINA (princess flower, glory bush)

Like most evergreens, this tall-growing semitropical shrub has its major growth in the spring, but its prime season for warm-climate gardeners is summer, when large purple flowers as much as 3 inches across blossom at the ends of its branches. It is a vigorous shrub, reaching heights up to 15

Cutting back · Shearing

UPRIGHT JAPANESE YEW
Taxus cuspidata capitata

Cutting back · Shearing

GIANT ARBORVITAE
Thuja plicata

LITTLE-LEAVED LINDEN
Tilia cordata

First pruning   Second pruning   Third pruning

CANADA HEMLOCK
Tsuga canadensis

Cutting back   Shearing

feet, but it may be pruned to a more compact size for growing in a greenhouse. Old, overgrown plants will readily sprout new branches after being cut back severely.

Prune tibouchina in the early spring before new growth starts, removing any damaged branches. Pinch back new growth lightly during the summer after flowering to encourage bushier growth and to control the plant's size.

TILIA (linden, ornamental lime)

Tall, rapid-growing, and naturally symmetrical in shape, lindens need almost no pruning, yet paradoxically they are favorite trees in Europe for that most drastic pruning method, pollarding. Linden trees develop dark, massive trunks and a large number of slender, lightweight branches that provide dense shade. In early summer they produce extremely fragrant pendulous clusters of tiny blossoms that are followed by small inedible berries. Prune linden trees in winter, when they are dormant.

For a strong shade tree, train a linden sapling to a central leader and cut back lateral branches by one half. As the tree develops, remove any low branches that emerge above the desired level of clearance and any branches that rub against each other or form weak crotches. Old lindens may need to be thinned at the crown to admit light and air to the interior.

TRAILING LANTANA See *Lantana*
TREE OF HEAVEN See *Ailanthus*
TRUMPET CREEPER See *Campsis*
TRUMPET FLOWER See *Campsis*
TRUMPET VINE See *Bignonia*

TSUGA (hemlock)

Graceful evergreen hemlocks with their long, sweeping branches require almost no pruning when they are grown as specimen trees. Without any shaping, fast-growing Canada hemlocks naturally form pyramids or columns; Sargent's weeping hemlocks form mounds of soft, pendulous branches; Japanese hemlock, a slow-growing variety, develops into a low bush with multiple trunks.

Hemlocks are easily sheared into hedges or screens, and since dormant buds on the old wood of these twiggy plants readily sprout new shoots, the plants can be sheared at almost any time. Allow young hedge plants to approach the desired size, pinching or cutting back any straggly branches in early spring before new growth begins. Then, to maintain a hedge at a given size, shear off the new growth in late spring or early summer, cropping it closely and gradually narrowing it toward the top. The outline of a hemlock hedge, with its soft, needle-leaved foliage, becomes feathery during the growing season; for a formal appearance, a light shearing in midsummer may be called for.

To rejuvenate a hemlock hedge that is overgrown, shear it heavily in early spring before new growth starts so that the new shoots will hide the cut ends of the branches.

TUPELO See *Nyssa*
TULIP POPLAR See *Liriodendron*
TULIP TREE See *Liriodendron*

U

ULMUS (elm)

Most of the disease-resistant elms brought into the United States to replace the American elm grow rapidly and may become too dense when mature. All elms, both tall species, such as smooth-leaved and Siberian elms, and the smaller Chinese elm should be pruned in late winter or early spring.

To train a young smooth-leaved elm, cut back its lateral branches until it is about 15 feet tall to force strength into the trunk. Then allow the tree to grow naturally, removing only damaged limbs and any that mar the vertical shape.

To train a naturally forking Oriental elm to a single trunk, select the strongest upright stem when the forking begins and cut away any other stems. Thin out some of the top branches while the tree is young to establish scaffold branches. As the tree ages, thin out the crown when it becomes overgrown, removing about one third of the older wood. Also remove any branches that have been damaged; these are particularly prevalent on the Siberian elm, whose wood is more brittle than that of other elms.

UMBRELLA TREE See *Catalpa* and *Melia*

 V

VACCINIUM (blueberry)

Properly pruned, a blueberry bush is decorative all year round, and pruning also improves the quantity and quality of its fruit. The bush is naturally upright with a rounded top; it blooms in spring and produces berries in early summer on old wood. In fall, its foliage and branches turn dark red.

At planting, cut young bushes back to half their height. Allow them to grow for four or five years, and then prune them in winter or early spring. Pruning for ornament is essentially the same as pruning for fruit production, although it needs to be practiced less faithfully. Canes bear the most fruit and the biggest fruit when they are eight to 10 years old and sparsely branched. To create this ideal condition, thin out some of the branches along the canes each year. When the canes become less productive, remove them entirely at ground level, to make room for young, vigorous canes. These same procedures improve the blueberry bush's appearance.

VIBURNUM

An extremely large group of highly ornamental shrubs, viburnums vary widely in size and shape. Some are tall and treelike and produce large flower clusters and bright berries; others are dense dwarfs with inconspicuous flowers. Most of them maintain an attractive appearance through the growing season and may be used as specimen plants and border plantings as well as hedges. Those that flower in the spring should be pruned after flowering since they developed the buds in the previous season, while summer-blooming species should be pruned in late winter or early spring. Dwarf viburnums grown as globe-shaped specimen plants or as formal hedges should be clipped or sheared in spring.

To train newly planted shrub viburnums, cut back the smaller canes to ground level and reduce the remaining canes to various lengths to establish a well-balanced framework; then thin out the weak shoots that appear at the base. Thereafter, remove deadwood, crowded branches and occasionally some of the oldest canes.

To train a larger viburnum as a small tree, select one stem as a central leader and cut back the other stems to ground level. Remove all lateral branches from the central leader up to 4 to 6 feet; then allow the branches to grow naturally, pruning them only to remove deadwood or overcrowded growth. Many newer viburnums sold as specimen plants are grafted stock and should be pruned accordingly. Do not prune below the graft, and be sure to remove suckers that sprout from below the graft as soon as they appear.

VIRGIN'S BOWER See *Clematis*
VIRGINIA CREEPER See *Parthenocissus*

CHINESE ELM
Ulmus parvifolia

First pruning · Second pruning · Third pruning

FRAGRANT SNOWBALL
Viburnum carlcephalum

Thinning · Cutting back · Shearing

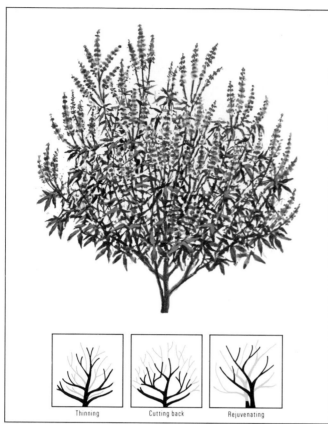

CHASTE TREE
Vitex agnus-castus

Thinning   Cutting back   Rejuvenating

Trellis training

HYBRID GRAPE
Vitis hybrid

VITEX (chaste tree)

This large bushy shrub bears summer-blooming flowers on growth of the current season. It is prone to winterkill in cold climates; if damage is severe, prune the plant in late winter, cutting back to 10 to 15 inches from the ground. If damage is slight, cut back to the base of the previous year's growth.

In milder climates, prune the chaste tree in early spring, before new growth starts. Cut back the stems to the base of the previous year's growth, leaving two or three healthy buds. When new growth starts, thin out weak or crowded shoots. Some gardeners prefer to prune after new growth starts, removing one third of the old wood at the ground and cutting back the remaining stems to half their length.

VITIS (grape)

Grape vines produce their fruit from canes that grew in the current season. If left unpruned, these smooth young branches become a fibrous-barked vine that in five years extends 30 feet or more from the main trunk and produces inferior fruit. The main purpose of pruning is to prevent this and to ensure an annual supply of year-old wood.

American grapes grown for fruit can be trained to a two-wire fence or trellis with a lower wire 2 or 3 feet from the ground and an upper wire 4 or 5 feet high. Pruning is done before sap flows in spring, while the vines are still dormant; otherwise the cuts bleed profusely. To train young plants, cut them back to two buds at the time of planting. In summer, after these buds have sprouted, remove the weaker of the two; stake the plant vertically. When the vine reaches the top wire, tie it to the wire and cut back the tip so it is about 3 inches above the wire. This forces the vine to produce side growth. The following winter, remove all but four strong side shoots, two for each wire. When these shoots develop about 10 buds, cut them back to about six buds. These buds will produce the first grape-bearing branches.

Thereafter, the object of pruning is to renew the fruiting branches. Cut back the previous year's fruit-bearing branch to two to four buds, depending on its vigor. Southern muscadine grapes are vigorous enough so you can permit the arms to grow 8 to 10 feet long with two or three bud spurs on each. European grapes are frequently grown by the head-spur system, whereby a strong trunk is created with a crown of fruit-bearing spurs at its top.

W

WALNUT See *Juglans*
WAX MYRTLE See *Myrica*

WEIGELA

Characteristically rounded or arching, with many stems and a profusion of flowers, the vigorous, fast-growing weigela shrubs need some annual pruning to keep them from becoming overgrown. They generally bloom in late spring on the previous season's growth, and should therefore be pruned as soon as the flowers fade, before seed heads form. Cut back the branches to healthy new shoots. Thin out weak or crowded canes and remove old canes at the ground. To rejuvenate an old or overgrown plant, cut back the entire shrub to the ground in early spring before new growth begins. As new shoots develop, thin out those that are weak or crowded. Remove dead or damaged branches or canes in the spring.

WHITEBEAM See *Sorbus*
WILLOW See *Salix*
WILSON PEARLBUSH See *Exochorda*
WINTERCREEPER See *Euonymus*

WISTERIA

A vigorous twining vine, wisteria develops gnarled, woody trunks as it grows older. It can be grown flat against a support, as a shrub, or trained as a freestanding tree. In late spring, cascading clusters of white, pink or violet-to-blue flowers open from bushy buds formed the year before.

For bigger, more luxuriant blooms, thin out old stems severely in spring and pinch the ends of trailing new growth throughout summer, and eliminate some of the current year's growth by cutting back to two or three buds in fall. Remove suckers at the base of the plant as they appear.

Rejuvenate old vines by cutting them to the ground, then selecting new shoots to train. To develop a standard or tree wisteria, prune a young plant back to a single stem and stake it for support until a trunk develops. The following year, pinch the top shoots back to one or two nodes and remove any side branches that sprout on the trunk. In subsequent years, continue cutting the branches back in the same way and removing any side branches from the trunk.

WITCH HAZEL See *Hamamelis*
WOODBINE See *Parthenocissus*

X

XANTHORIZA (yellowroot)

Characterized by its yellow bark when young and, as the common name implies, its yellow roots, this low-growing deciduous shrub is used primarily as a ground cover for damp, shady locations. It forms dense mounds of pale green celery-like foliage up to 2 feet high on branchless stems. In spring it produces inconspicuous clusters of purple-brown flowers. Spreading rapidly by means of underground stems, it may need root pruning to retard it.

Prune yellowroot in late winter before new growth starts. Thin out dead or damaged branches and cut back those that spoil the plant's symmetry.

Y

YEDDA HAWTHORN See *Raphiolepis*
YELLOW POPLAR See *Liriodendron*
YELLOWROOT See *Xanthoriza*
YEW See *Taxus*
YEW PODOCARPUS See *Podocarpus*

Z

ZELKOVA

The zelkova tree, because it has a similar leaf pattern, is often planted as a substitute for the disease-prone American elm. With regular pruning, it can indeed be made to resemble the elm's vase shape and its open, spreading crown. But its normal branching pattern is lateral, especially when it is young, and its natural tendency is to develop a short trunk with many upright branches.

Prune zelkovas in late summer or early fall, when growth has slowed and there is less likelihood that the tree will send up suckers. During this period, it is easier to see where the growth needs to be opened to show the tree's interesting branching structure. To train a young tree to a single trunk, select one stem as a central leader and remove all others. Gradually remove side branches below a height of 8 to 10 feet; this may take several years but it will result in a vertical growth pattern and a gigantic tree that may at maturity reach a height of 80 feet or more.

Zelkova can also be trained to a bushier tree with several trunks and a rounded crown; or it can be dwarfed by planting it in a container and constantly pinching it back.

Thinning   Cutting back   Rejuvenating

WEIGELA
Weigela 'Vanicek'

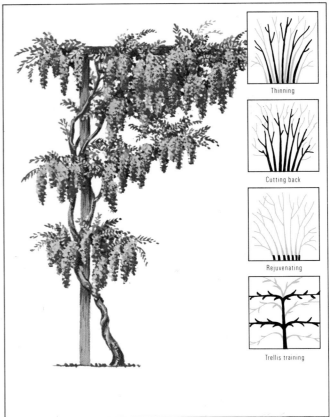

Thinning

Cutting back

Rejuvenating

Trellis training

CHINESE WISTERIA
Wisteria sinensis

Bibliography

Atkinson, Robert E., *Dwarf Fruit Trees Indoors and Outdoors*. Van Nostrand Reinhold, 1972.

Ballard, Ernesta Drinker, *The Art of Training Plants*. Barnes & Noble, 1962.

Barber, Peter, *The Trees Around Us*. Chicago University Press, 1975.

Baumgardt, John Philip, *How to Prune Almost Everything*. William Morrow & Co., Inc., 1968.

Brockman, C. Frank, *Trees of North America*. Western Publishing Co., Inc., 1968.

Brooklyn Botanic Garden, *Pruning Handbook*. BBG, 1966.

Brooklyn Botanic Garden, *Trained and Sculptured Plants*. BBG, 1961.

Brown, George E., *The Pruning of Trees, Shrubs and Conifers*. Winchester Press, 1972.

Chidamian, Claude, *The Book of Cacti and Other Succulents*. Doubleday, 1958.

Chittenden, Fred J., ed., *The Royal Horticultural Society Dictionary of Gardening*, 2nd ed. Clarendon Press, 1974.

Christopher, Everett P., *The Pruning Manual*. The Macmillan Co., 1954.

Cloud, Katharine M-P, *Evergreens For Every State*. Chilton Co., 1960.

Collingwood, G. H., and Brush, Warren D., *Knowing Your Trees*. The American Forestry Association, 1964.

Curtis, Charles H., and Gibson, W., *The Book of Topiary*. Bodley Head, London, 1904.

Dietz, Marjorie J., *Concise Encyclopedia of Favorite Flowering Shrubs*. Doubleday & Co., Inc., 1963.

Evans, Charles M., *New Plants From Old*. Random House, 1976.

Faust, Joan L., ed., *The New York Times Garden Book*. Alfred A. Knopf, Inc., 1973.

Fleming, John, and Honour, Hugh, *Dictionary of the Decorative Arts*. Harper and Row, 1977.

Foster, Catharine Osgood, *Plants-A-Plenty*. Rodale Press, 1977.

Free, Montague, *Plant Pruning in Pictures*. Doubleday & Co., Inc., 1961.

Garner, R. J., *The Grafter's Handbook*, rev. ed. Faber and Faber Ltd., London, 1967.

Griffin, Appleton P. C., *A Catalogue of the Washington Collection in the Boston Athenaeum*. The Boston Athenaeum, 1897.

Grounds, Roger, *The Complete Handbook of Pruning*. Ward Lock Ltd., London, 1973.

Grounds, Roger, *Practical Pruning*. Ward Lock Ltd., 1974.

Hadfield, Miles, *Topiary and Ornamental Hedges*. Adam & Charles Black, London, 1971.

Hartmann, Hudson, and Kester, Dale E., *Plant Propagation, Principles and Practices*. Prentice-Hall, Inc., 1975.

Hill, Lewis, *Fruits and Berries for the Home Garden*. Alfred A. Knopf, 1977.

Hoobler, Dorothy and Thomas, *Pruning*. Grosset & Dunlap, 1975.

Hottes, Alfred C., *How to Increase Plants*. De La Mare Co., 1949.

Houghton, A. D., *The Cactus Book*. The Macmillan Co., 1930.

Howell, O. B., *Pruning Without Pain*. Montana State University Press, 1958.

Hudson, Roy L., *The Pruning Handbook*. Prentice-Hall, Inc., 1972.

Ishimoto, Tatsuo and Kiyoko, *The Art of Shaping Shrubs, Trees and Other Plants*. Crown Publishers, Inc., 1966.

Jackson, Donald, and Twohig, Dorothy, eds., *The Diaries of George Washington*. University Press of Virginia, 1978.

Kraft, Ken and Pat, *Grow Your Own Dwarf Fruit Trees*. Walker, 1974.

Kramer, Jack, *Gardening and Home Landscaping Guide*. Arco Publishing Co., 1968.

Lemmon, Robert S., *The Best Loved Trees of America*. Literary Guild of America, Inc., 1946.

Lucas, I. B., *Dwarf Fruit Trees for Home Gardens*. Dover, 1977.

McCurdy, Dwight R., Spangenberg, William Greg, and Doty, Charles Paul, *How to Choose Your Tree*. Southern Illinois University Press, 1972.

Maino, Evelyn, and Howard, Frances, *Ornamental Trees: An Illustrated Guide to Their Selection and Care*. University of California Press, 1955.

Marquand, Allan, *Luca Della Robbia*. Hacker Art Books, 1972.

Miller, Philip, *The Abridgement of the Gardeners Dictionary*, 6th ed. John and Francis Rivington, London, 1771.

Murphy, Richard C., and Meyer, William E., *The Care and Feeding of Trees*. Crown Publishers, Inc., 1969.

Perkins, Harold O., *Espaliers and Vines for the Home Gardener*. D. Van Nostrand Co., Inc., 1964.

Pierot, Suzanne, *The Ivy Book*. The Macmillan Co., 1974.

Pirone, P. P., *Tree Maintenance*. Oxford University Press, 1972.

Powell, Thomas and Betty, *The Avant Gardener*. Houghton Mifflin, 1975.

Rehder, Alfred, *Manual of Cultivated Trees and Shrubs*, 2nd ed. The Macmillan Co., 1940.

Smith, Alice Upham, *Trees in a Winter Landscape*. Holt, Rinehart and Winston, 1969.

Southwick, Lawrence, *Dwarf Fruit Trees*. Garden Way, 1972.

Staff of the L. H. Bailey Hortorium, Cornell University, *Hortus Third: A Dictionary of Plants Cultivated in the United States and Canada*. Macmillan Publishing Co., 1976.

Steffek, Edwin F., *Pruning Made Easy*. Holt, Rinehart and Winston, 1962.

Steffek, Edwin F., *The Pruning Manual*. Little, Brown & Co., 1969.

Stevenson, Tom, *Pruning Guide for Trees, Shrubs and Vines*. Robert Luce, 1964.

Sunset Editors, *Basic Gardening Illustrated*. Lane Publishing Co., 1976.

Sunset Editors, *Pruning Handbook*. Lane Publishing Co., 1972.

Sunset Editors, *Succulents and Cactus*. Lane Publishing Co., 1970.

Sunset Editors, *Western Garden Book*. Lane Publishing Co., 1976.

Symonds, George W. D., *The Tree Identification Book*. William Morrow & Co., Inc., 1958.

Taylor, Norman, *The Guide to Garden Shrubs and Trees*. Houghton Mifflin Co., 1965.

Taylor, Norman, ed., *Taylor's Encyclopedia of Gardening*, rev. ed. Houghton Mifflin Co., 1961.

Tukey, Harold Bradford, *Dwarfed Fruit Trees*. The Macmillan Co., 1964.

U.S. Dept. of Agriculture, *Bridge Grafting and Inarching Damaged Fruit Trees*. U.S. Government Printing Office, 1968.

U.S.D.A. Agricultural Research Service, *High-density apple orchards—planning, training, and pruning*. U.S. Government Printing Office, 1975.

Watkins, John V., and Sheehan, Thomas J., *Florida Landscape Plants: Native and Exotic*, rev. ed. University Presses of Florida, 1975.

Weiner, Michael, *Plant a Tree: A Working Guide to Regreening America*. Macmillan Publishing Co., Inc., 1975.

Wittrock, Gustave L., *The Pruning Book*. Rodale Press, Inc., 1971.

Wright, R. C. M., *The Complete Handbook of Plant Propaga-*

tion. Macmillan Publishing Co., Inc., 1975.

Wyman, Donald, *Shrubs and Vines for American Gardens.* Macmillan Publishing Co., Inc., 1969.

Wyman, Donald, *Trees for American Gardens.* The Macmillan Co., 1965.

Wyman, Donald, *Wyman's Gardening Encyclopedia.* Macmillan Publishing Co., Inc., 1977.

Yang, Linda, *The Terrace Gardener's Handbook.* Doubleday, 1975.

Yepsen, Roger B., *Trees for the Yard, Orchard, and Woodlot.* Rodale Press, 1976.

Zion, Robert L., *Trees for Architecture and the Landscape.* Reinhold Book Corp., 1968.

Zucker, Isabel, *Flowering Shrubs.* D. Van Nostrand Co., Inc., 1966.

Acknowledgments

The index for this book was prepared by Anita R. Beckerman. For their help in the preparation of this book, the editors wish to thank the following: American Horticultural Society, Mount Vernon, Va.; Mai K. Arbegast, Berkeley, Calif.; J. Roy W. Barrette, Amen Farm, Brooklin, Me.; Mr. and Mrs. Charles T. Berry Jr., Upperville, Va.; Mr. F. Joseph Carstens, Horticulturist, Longwood Gardens, Kennett Square, Pa.; Mr. and Mrs. Allen Charlin, Beverly Hills, Calif.; Stelvio Coggiatti, President, Garden Club, Rome, Italy; Mr. and Mrs. Harry A. Councilor, Alexandria, Va.; Dr. John Creech, Director, National Arboretum, Washington, D.C.; Mrs. Lockwood de Forest, Santa Barbara, Calif.; James Draper, Associate Curator, Department of European Sculpture and Decorative Arts, Metropolitan Museum of Art, New York City; Mr. and Mrs. John W. Drayton, Paoli, Pa.; Dr. T. R. Dudley, Research Botanist, U.S. National Arboretum, Washington, D.C.; Frances Evelyn Eiger, Hillsborough, Calif.; Professor Carra Ferguson, Department of Fine Arts, Georgetown University, Washington, D.C.; The Filoli Center, Woodside, Calif.; Galper Baldon Associates, Venice, Calif.; Professor Owen Gingrich, Harvard-Smithsonian Center for Astrophysics, Cambridge, Mass.; Betty and William Grau, Alexandria, Va.; Guardian Tree Experts, Inc., Alexandria, Va.; Grace Hall, Thomas D. Church and Associates, San Francisco, Calif.; Dr. and Mrs. George C. Henny, Philadelphia, Pa.; Ron Herman, Berkeley, Calif.; M. T. Hirschkoff, Paris, France; Dr. and Mrs. John E. Hopkins, Wayne, Pa.; Wanda Jablonski, New York City; Mrs. Sydney Keith, Chestnut Hill, Pa.; Art Klempner, Birnam Wood Golf Club, Santa Barbara, Calif.; Henry Leuthardt, East Moriches, N.Y.; Jessie McNab, Associate Curator, Department of European Sculpture and Decorative Arts, Metropolitan Museum of Art, New York City; Mrs. John Maury, Washington, D.C.; Robert Moore's Topiary, Anaheim, Calif.; The Mount Vernon Ladies' Association of the Union, Mount Vernon, Va.; J. Liddon Pennock Jr., Meadowbrook Farm Greenhouse, Meadowbrook, Pa.; Mr. and Mrs. Herbert Peters, Montecido, Calif.; Edward C. Plyler, Alexandria, Va.; Sally Reath, Devon, Pa.; P. Stageman, Librarian, Royal Horticultural Society, London, England; Dr. William Louis Stern, Professor of Botany, University of Maryland, College Park, Md.; Charles D. Webster, President, The Horticultural Society of New York, New York City; Weston Nurseries, Hopkinton, Mass.; Mrs. Jean Wolff, San Francisco, Calif.; Mrs. Frank L. Wright, Alexandria, Va.

Picture credits

The sources for the illustrations in this book are shown below. Credits from left to right are separated by semicolons, from top to bottom by dashes. Cover: Bernard Askienazy. 4: Frederick R. Allen. 6: Tom Tracy. 10, 11: Drawings by Susan M. Johnston. 12 through 19: Drawings by Kathy Rebeiz. 21: Tom Tracy. 22: Bernard Askienazy. 23, 24, 25: John Neubauer. 26, 27: Bernard Askienazy. 28: Tom Tracy. 29, 30, 31: Bernard Askienazy. 32: John Neubauer. 34 through 39: Drawings by Kathy Rebeiz. 40, 41: Drawings by Susan M. Johnston. 42 through 49: Drawings by Kathy Rebeiz. 53: John Neubauer. 54: Bernard Askienazy; John Neubauer. 55: Richard Weymouth Brooks, except bottom right, John Neubauer. 56: Sonja Bullaty and Angelo Lomeo, courtesy Weston Nurseries; John Neubauer. 57: Bernard Askienazy. 58: Scala, courtesy Victoria and Albert Museum, London, Crown Copyright. 63 through 70: Drawings by Kathy Rebeiz. 74: General Research and Humanities Division, The New York Public Library, Astor, Lenox and Tilden Foundations. 79 through 82: John Neubauer. 86, 87: Drawings by Susan M. Johnston. 88: Drawings by Kathy Rebeiz. 91: Tom Tracy, topiary designed by Robert Moore. 92: John Neubauer. 93: Tom Tracy. 94, 95: Richard Weymouth Brooks. 96: John Neubauer. 97: Wolf von dem Bussche. 98, 99: John Neubauer. 100 through 153: Artists for encyclopedia illustrations listed in alphabetical order: Adolph E. Brotman, Richard Crist, Mary Kellner, Gwen Leighton, Harry McNaught, Rebecca Merrilees, John Murphy, Kathy Rebeiz, Eduardo Salgado and Barbara Wolff.

Index

Numerals in italics indicate an illustration of the subject mentioned.

PRINTED IN U.S.A.